The Compleat Lemon

HOLT, RINEHART AND WINSTON • NEW YORK

Published by Holt, Rinehart and Winston, 383 Madison
Avenue, New York, New York 10017.
Published simultaneously in Canada by Holt, Rinehart
and Winston of Canada, Limited.

Library of Congress Cataloging in Publication Data
Casson, Chris.
The compleat lemon.
Includes index.
1. Cookery (Lemons) I. Lee, Susan, 1943–
joint author. II. Title.
TX813.L4C37 641.6′4′334 78–18261
ISBN 0–03–041881–X

First Edition

Illustrations by Kimble Mead

Designer: Amy Hill
Printed in the United States of America
10 9 8 7 6 5 4 3 2 1

For Kevin, Peter,
Max, and Chico

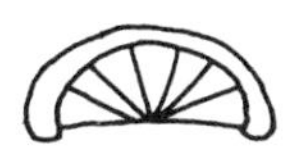

"A Persian's heaven is easily made: Tis but black eyes and lemonade."

Intercepted Letters
Thomas More

"My being in Yorkshire was so far out of the way, that it was actually twelve miles from a lemon."

Sydney Smith

Contents

Acknowledgments

Our special thanks to Marie-Claude Butler, Ann Marie Casson, Pat Casson, Tom Casson, Ray Cole, Don Cutler, B. J. Doerfling, Cal Fentress, Nancy Forbes, Kathy Johnson, Mary Kay Johnson, Amy Hill, Leola MacDonald, Connie Madden, Sally Madden, Mary Madden, Dan Mahan, Bridget Marmion, Ellyn Polshek, Jane Price, Arthur Rosenblatt, Jay Schreick, Reva Schwartz, Patsy Sellew, Irene Skolnick, Anice Smelin, Emmy Smith, Amanda Vaill, Tom Victor, and Tina Yatrakis.

Introduction

August Escoffier put it in his sauce allemande. Irma Rombauer dropped it in her espresso. Michael Field squeezed it on his chicken. Marcella Hazan mashed it with capers. Fernand Point dribbled it on his pastry.

What?

Lemon. Of course. Lemon has been prized as a taste in all major cuisines since the first Indian doused a dish of lentils with its juice thousands of years ago. Americans alone consume, in one way or another, more than four billion lemons annually—about five pounds per mouth.

For, although its role as a seasoning makes it as ubiquitous as salt, it is also a necessary element in hundreds of dishes. Lemon lends its essence to recipes both as simple as lemonade and as complicated as lemon soufflé; from straightforward sole with lemon to fancy French hollandaise; from robust Greek egg-lemon soup to subtle Cantonese lemon chicken; from hors d'oeuvres like oysters with lemon to desserts like tarte au citron.

Lemons have had not only a versatile culinary history but a rather long one as well. Nobody knows for sure just where and when lemons were first cultivated. Probably in India. But lemons have been grown for centuries in Southeast Asia, North Africa, and Mediterranean countries such as Italy, Egypt, and Spain. Lemons arrived in North America with Columbus, who brought some seeds along on his explorations. Cultivation took root in Mexico, Florida, and later, California. Today almost all lemons grown in America are grown in California.

California lemons are not, of course, ripened on the tree. Shipping lemons from California to Vermont means lemons must be picked when they're still green. As lemons ripen, they progress from green, with a high acid content, to yellow, with lower acid, so color is a fine indicator of readiness. Don't buy lemons unless they are perfectly and gloriously yellow. Once bought, lemons should last—in the refrigerator or a cool place—anywhere from thirty to sixty days.

Lemons have two usable ingredients for cooking: the juice and the yellow part of the peel, called zest. There is more than one way to juice a lemon; sides to the controversy range from the simple roll-it-around-a-countertop to the elaborate drop-it-in-boiling-water-for-a-minute. All procedures are designed to break down the membrane compartments. Our own opinion is that a satisfactory amount of juice can be obtained by pressing the lemon between the palms of your hands several times (an isometric approach which, incidentally, aids in the development of pectorals) before cutting and juicing. One medium lemon should yield 1½ to 2 tablespoons of juice. A good rule of thumb is to figure six lemons will make 1 cup of juice.

The zest is simply the peel without any of the bitter white pulp between it and the inside of the lemon. Lemon zest is used frequently in baking because it lends an intense lemon flavor without adding any liquid or much dry ingredient to the recipe. The robust flavor of zest comes from the high concentration of lemon oil in the peel.

Lemons have lots of uses other than for flavor in cooking and baking. First of all, they're incredibly decorative. A bunch of lemons in an Imari bowl or on a blue faïence plate make a handsome still-life arrangement. A large wooden salad bowl heaped with lemons is a charming centerpiece. We even like to fill a silver porringer with lemons which have been pricked in a few places with a pin so they exude a fresh lemon scent.

Also, lemons can be employed in cleaning all sorts of things. A few drops of lemon juice in the rinse water—as the commercials might say—causes linen handkerchiefs to look like fresh-fallen snow and makes windows as transparent as the gardol shield. Lemon juice can also clean people; it is thought to remove freckles, make hair shiny, and tighten pores when mixed in facial masks. Squeezing lemon juice on cut fruit and vegetables prevents them from turning brown. And the oil of lemons has caused countless grand pianos and dining room sideboards to gleam.

Lemons are a health food, too. Centuries ago lemons were used to prevent scurvy because of their high vitamin C content—one lemon provides almost 90 percent of an adult's daily requirements. Today, people on low-salt or salt-free diets dribble lemon juice on their food as a sodium substitute. For people on weight-losing diets, using lemon juice instead of salad dressing (or whatever first aid required to make low-calorie food palatable) adds only four calories per tablespoon. There is also considerable opinion which holds that lemon zest improves digestion.

Lemons have earned both their long history and present popularity. They are delicious, beautiful, and useful. In fact, lemons are even inspirational; after all, when Proust crumbled madeleine in his tea, it was probably the scent of lemon in the madeleine that undid him. . . .

The recipes that follow celebrate the lemon as a premier ingredient. Some of the recipes are obvious, others are subtle, some even astonishing in their use of lemon. No matter, all depend on The Lemon. So be fair: use only fresh lemon. Squirt fresh juice and grate fresh zest. We promise you'll be enchanted and rewarded.

NUTRIENTS IN THE LEMON

Here are all the good things inside one medium-size lemon, approx. 2½-inch diameter.

Weight	110 grams
Water	87.4%
Food Energy	22 calories
Protein	1.3 grams
Fat	.3 grams
Carbohydrate	11.7 grams
Calcium	66 mg.
Phosphorus	16 mg.
Iron	.8 mg.
Sodium	3 mg.
Potassium	158 mg.
Vitamin A	30 IU (International Units)
Thiamine	.05 mg.
Riboflavin	.04 mg.
Niacin	.2 mg.
Vitamin C	84 mg.

Hors d'Oeuvres

HORS D'OEUVRES

The first two recipes in this section, Oysters on the Half Shell and Fresh Caviar, make a simple and classic—if expensive—beginning for any dinner party. Moreover, these recipes are a perfect start for a cookbook celebrating the lemon as the often forgotten underpinning of flavor. Lemon is as crucial as the oysters or the caviar in these recipes. As good as the oysters or caviar are, either one without a few drops of lemon is unimaginable.

Tapenade is a Mediterranean dish; its blend of olives, tuna, and anchovy would be overpoweringly salty and fishy if not cut with lemon juice. This is also true for the anchovies in the Anchoyade.

The identical marinade and the marinating process used in the Mushrooms à la Grecque can be used with other vegetables such as cucumbers or green peppers—for cold hors d'oeuvres.

And last, the Cold Purée of Eggplant with Lemon Juice, the Taramasalata, and the Hummus are great (and very correct) when served with wedges of warmed pita bread.

OYSTERS ON THE HALF SHELL

Serves 6

30 oysters, well scrubbed
3 **lemons**

Shucking oysters is rather difficult for the uninitiated, so you might wish to buy oysters which have already been opened. If not, prepare yourself with an oyster knife and hold the oyster, smaller half up, in your palm. Insert the knife into the hinge, turning and prying until you have cut through the hinge muscle. Then run the knife around the rest of the shell, open, and discard top (smaller) shell.

Arrange five oysters over cracked ice on each of six individual serving plates. Decorate each plate with half a lemon, cut into wedges. Serve.

FRESH CAVIAR

Serves 6

8 to 10 ounces fresh beluga caviar (depending on how much you like your guests)
1 **lemon,** cut into 6 wedges

Mold the caviar into a mold in a small glass bowl. Place the glass bowl in a larger bowl filled with cracked ice. Give each guest a plate, a lemon wedge, and a tiny wooden or other nonmetal spoon. Bring on the caviar.

MELON AND COLD HAM

Serves 6

6 thin slices cold Westphalian ham, prosciutto,
    or first-rate cooked ham
6 slices cantaloupe or Persian melon, 1 inch thick,
    rind removed
12 Mediterranean-style black olives
Freshly ground black pepper
2 **lemons,** quartered

Loosely roll each slice of ham into a cigar shape and place down the center of a long chilled dish. Arrange the melon slices in an overlapping pattern around the sides of the dish. Scatter the olives over all. Dust the ham and melon with the pepper. Wedge the lemon quarters between slices of melon. Serve chilled.

ANCHOYADE

Serves 4 to 6

Two 2-ounce cans flat anchovy fillets, undrained
2 garlic cloves, minced
2 teaspoons tomato paste
1 tablespoon olive oil
3 tablespoons **lemon** juice
Freshly ground black pepper
12 slices French bread, ½ inch thick
24 pimiento strips

Preheat oven to 500 degrees. Put the anchovies, garlic, and tomato paste in a bowl. Mash, using the back of a large spoon, until very smooth. Pour in the oil, slowly, beating vigorously. Slowly pour in the lemon juice, still beating. Add pepper to taste.

Toast the bread on one side under broiler for 4 minutes. Remove the slices and spread the untoasted side with anchovy mixture. Cross each slice with two strips of pimiento and then run them under the broiler for 4 minutes. Serve immediately.

TAPENADE

Serves 8

3½-ounce can tuna, oil-packed and undrained
3¼-ounce can large black olives, pitted
2-ounce can anchovy fillets with capers
2 tablespoons olive oil
¼ cup **lemon** juice
Few drops of Cognac, to taste
Freshly ground black pepper to taste

Mash the tuna, olives, and anchovies all together until you have a dense paste (or put everything into a food processor). Slowly add the olive oil, by droplets, beating vigorously. Then slowly add the lemon juice, still beating. At the end, beat in a few drops of Cognac and the pepper. Chill for 1 hour or more before serving. A superb spread for a good French bread.

GUACAMOLE

Serves 6

2 ripe avocados
1 ripe tomato, peeled, seeded, and chopped
1 scallion, minced
½ teaspoon chili powder
2 teaspoons olive oil
1 tablespoon **lemon** juice
½ teaspoon ground coriander
Salt and freshly ground black pepper

Cut the avocados in half and remove the seeds. Pare away the skin, then chop coarsely. Put the avocado chunks in a mixing bowl and, using the back of a wooden spoon, mash to a smooth paste. Add the tomato, scallion, chili powder, olive oil, lemon juice, and coriander, and mash until mixture is a smooth paste. Add salt and pepper to taste. Refrigerate for at least 1 hour and serve chilled. Use Guacamole as a dip for almost anything crispy—from chips to raw vegetables.

MUSHROOMS À LA GRECQUE

Serves 6

5 cups water
1/4 cup olive oil
1/2 cup **lemon** juice
1/4 cup chopped scallions
5 sprigs parsley
1 celery stalk
1/2 teaspoon dried thyme
10 black peppercorns
1 teaspoon salt
1 pound fresh mushrooms

Simmer all ingredients except mushrooms in a large heavy saucepan, covered, for 20 minutes.

Wash the mushrooms carefully. Add them to the simmering liquid, cover, and simmer for 10 minutes. Remove the mushrooms with a slotted spoon to a serving dish. Vigorously boil the marinade until it reduces to 1/4 cup. Strain the liquid and pour over the mushrooms. Serve at room temperature or slightly chilled.

EGGS STUFFED WITH SARDINES

Serves 6

6 large eggs, hard-boiled and shelled
4-ounce can sardines, drained
½ cup good commercial mayonnaise
2 tablespoons **lemon** juice
2 teaspoons prepared Dijon-style mustard
¼ teaspoon salt
¼ cup minced fresh parsley
¼ cup minced fresh chives

Slice the eggs in half lengthwise. Carefully remove the yolks. Place yolks in a small mixing bowl, add the sardines, and mash with a fork until mixture is a smooth paste. Vigorously beat in the mayonnaise, lemon juice, mustard, salt, and parsley. You should have a lumpless purée. Taste for seasoning.

Using a teaspoon, fill the egg halves with the purée, making a little mound on the top. Put the eggs on a serving dish and scatter the chives over all.

LEMON EGG SPREAD

Serves 10 to 12

8 hard-boiled eggs
½ cup fresh mayonnaise
6 tablespoons **lemon** juice
1 garlic clove, finely minced
2 tablespoons finely chopped scallions
½ teaspoon dried dill weed
⅛ teaspoon dried thyme
⅛ teaspoon dried tarragon
1 drop hot sauce (tabasco-type sauce)
Salt and black pepper
Toast points
Parsley sprigs

Chop eggs coarsely in a large bowl. Mix in mayonnaise until well blended. Add the lemon juice, garlic, scallions, dill, thyme, tarragon, and hot sauce, and stir until smooth. Salt and pepper to taste. Serve on a platter with toast points, and scatter parsley over all.

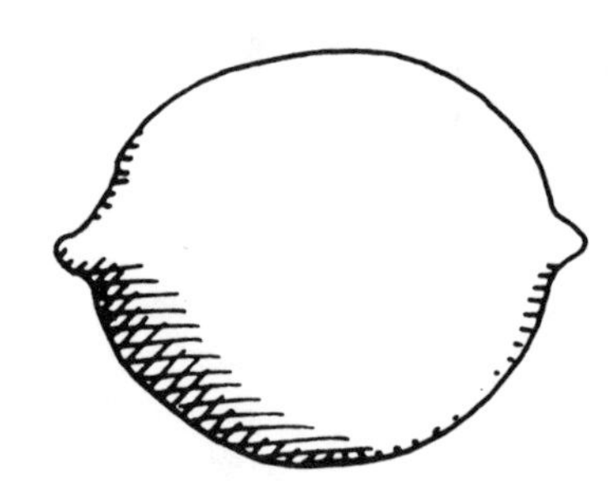

CEVICHE

Serves 6 to 8

1 pound sea scallops, cut in bite-size pieces
1 pound fillet of sole, cut in bite-size pieces
$\frac{3}{4}$ cup **lemon** juice
1 teaspoon salt
1 medium onion, sliced thin
3 tablespoons chopped fresh parsley

Place the scallops and sole in a deep bowl. Combine the lemon juice and salt, then pour over fish. Scatter the onion over all. Refrigerate, stirring occasionally, for 3 to 4 hours, or until fish has turned white and shrunk slightly. Sprinkle with parsley. Serve chilled with toothpicks to skewer the fish.

COLD PURÉE OF EGGPLANT WITH LEMON JUICE

Serves 6

1 medium eggplant
$\frac{1}{4}$ cup **lemon** juice
3 garlic cloves, crushed
$\frac{1}{2}$ tablespoon olive oil
$\frac{1}{4}$ cup minced onions
2 tablespoons minced fresh parsley
2 tablespoons chopped fresh basil (optional)
Salt and black pepper to taste

Preheat oven to 400 degrees. Prick the eggplant in four or five places with a fork, then roast it in the oven until it is soft, about 45 minutes. After eggplant has cooled, slice it open and scoop out the flesh. Discard the skin. In a medium-size bowl, mix together the eggplant, lemon juice, garlic, oil, onions, parsley, basil, salt, and pepper. Chill for 1 hour.

DILLY BEANS

Serves 8

2 pounds fresh string beans, trimmed
¾ cup vegetable oil
¼ cup wine vinegar
2 tablespoons **lemon** juice
1 cup water
1 tablespoon chopped fresh dill weed
1 garlic clove, peeled
¼ teaspoon crushed dried red pepper
1 teaspoon sugar
Salt to taste

Using a large kettle, bring 6 to 8 cups of water to boil. Add the string beans, cover, and simmer until beans are tender, about 15 minutes. Arrange the beans, vertically, in a quart-size jar. Combine the remaining ingredients to make the dill marinade. Pour the marinade into the jar and cover. Refrigerate for at least 24 hours—the longer the better—and serve chilled.

TARAMASALATA

Serves 6 to 8

5½ slices homemade-type white bread
1 cup water
½ cup tarama (carp roe)
¼ cup **lemon** juice
¼ cup grated onions
1 garlic clove, crushed
¾ cup olive oil

Cut the crusts off the bread and soak it in the water for 10 minutes. Squeeze it dry with your hands. Using the back of a large spoon, mash the bread until it is smooth. Add the tarama, a teaspoonful at a time, mashing the mixture smooth after each spoonful. Mash in the lemon juice and then the onions the same way. Add the garlic and mash again. Beat in the oil by droplets until mixture is smooth and thick. Refrigerate for at least 30 minutes before serving.

HUMMUS

Serves 6

2½ cups canned chickpeas
3 tablespoons light sesame oil or olive oil
¼ cup **lemon** juice
2 garlic cloves, crushed
½ teaspoon salt
2 tablespoons chopped fresh parsley
1 **lemon,** thinly sliced

Rub chickpeas through a strainer into a bowl. Alternately add oil and lemon juice, 1 tablespoon at a time, until oil is blended. Add the garlic and salt, then stir until paste is thick and smooth.

Refrigerate for 3 hours. Sprinkle with parsley and decorate with lemon slices.

SOUP
VIN

SOUP

Lemon and soup don't sound as though they would be particularly suited to each other. Of course, there is that clichéd consommé madrilène, flavored with lemon zest when hot and garnished with lemon wedges when cold and jelled; but what else? Actually, quite a bit else, as the following recipes show.

We begin with two traditional national soups: Greek Avgolemono and Mexican Gazpacho. Both are nourishing soups which can be served as the main course for lunch or as the substantial course for a light dinner.

The more exotic Pumpkin Soup is an excellent chilly-weather dish; Fresh Fruit Soup and Mussel Soup would be perfect for spring or summer menus.

Beet Soup with Dill constitutes a variation on borscht. Serve it hot or cold, but serve it authentically, with beer and pumpernickel.

AVGOLEMONO

Serves 4 to 6

6 cups chicken stock
¼ cup uncooked long-grain rice
4 eggs
Scant ¼ cup **lemon** juice
Salt and freshly ground black pepper
1 tablespoon chopped fresh mint (optional)

Bring the chicken stock to boil over high heat in a large saucepan. Add the rice, lower heat, and simmer, partially covered, for 10 to 15 minutes—rice should be firm, not mushy. Reduce heat. Meanwhile beat the eggs with whisk until frothy. Beat the lemon juice into the eggs. Stir in about ¼ cup of the chicken stock. Then slowly pour this egg mixture back into saucepan of chicken stock, stirring all the while. Heat without boiling for 5 minutes, or until soup thickens enough to coat spoon lightly. Be very careful not to boil soup or the egg yolks will curdle. Add salt, and pepper to taste, and scatter mint over soup.

GAZPACHO

Serves 8

SOUP

6 large ripe tomatoes, peeled and coarsely chopped
¼ cucumber, peeled, seeded, and chopped
½ cup chopped green pepper
½ teaspoon minced garlic
½ cup olive oil
¼ cup **lemon** juice
2 cups chicken stock
1 cup tomato juice
1 cup finely chopped onions
2 teaspoons salt
Freshly ground black pepper to taste
½ teaspoon paprika
1 teaspoon chopped fresh basil
1 teaspoon chopped fresh tarragon
2 teaspoons chopped fresh parsley

GARNISHES

1 cup croutons
1 cup chopped tomatoes
1 cup chopped green peppers
1 cup chopped onions
1 cup chopped cucumbers

Put first four soup ingredients—tomatoes, cucumber, green pepper, and garlic—in blender or food processor until smooth. Pour mixture into a large bowl and add olive oil, lemon juice, chicken stock, and tomato juice. Stir in remaining soup ingredients: onions, salt, pepper, paprika, basil, tarragon, and parsley. Mix until well blended, and refrigerate for 4 hours. Serve in very cold soup cups. Put garnishes of croutons, tomatoes, peppers, onions, and cucumbers in individual bowls so guests can help themselves.

MUSSEL SOUP

Serves 4

2 dozen mussels
1/3 cup olive oil
1/3 cup finely chopped onions
1/3 cup finely chopped celery
1 teaspoon minced garlic
3/4 cup dry white wine
2 teaspoons **lemon** juice
1 tablespoon finely chopped fresh basil
2 cups canned Italian tomatoes, chopped
1/2 teaspoon salt
1/2 teaspoon black pepper
1 tablespoon freshly grated **lemon** zest

Clean the mussels and remove the beards. Run them under cold water and set aside. Heat the olive oil in a large saucepan, add the onions, celery, and garlic, and stir over moderate heat for 10 minutes. Pour in the wine and bring mixture to a boil. Continue a high boil until mixture reduces to approximately 1/4 cup. Slowly stir in the lemon juice, basil, tomatoes and their juice, salt, and pepper, and let mixture simmer uncovered for 25 minutes. The mussels should now be added and cooked, covered, over high heat. Occasionally shake the pan to cook the mussels evenly. They should open in about 10 minutes. If not, cook a few minutes more. To serve, ladle mussels and juice into soup bowls. Sprinkle lemon zest over all.

PUMPKIN SOUP

Serves 6

1 tablespoon butter
2 tablespoons chopped onion
2½ cups cooked pumpkin
2 tablespoons **lemon** juice
2 cups milk
2 tablespoons brown sugar
½ teaspoon ground nutmeg
½ teaspoon ground cinnamon
½ teaspoon salt
Freshly ground black pepper to taste
2 cups chicken stock
¼ cup light cream
6 long strips of **lemon** zest

Melt the butter in a large heavy saucepan. Add the onion and cook over moderate heat until transparent—about 3 minutes. Stir in the pumpkin, lemon juice, milk, brown sugar, nutmeg, cinnamon, salt, pepper, and stock. Cook over moderate heat for 10 minutes. Remove from heat, let cool to room temperature, and force mixture through a fine sieve or use a food processor. Return soup to stove, stir in light cream, and cook over moderate heat for 15 minutes. Ladle into soup bowls and float a lemon zest strip in each bowl.

BEET SOUP WITH DILL

Serves 6

1½ pounds fresh beets
1 teaspoon salt
5 cups beef broth
2 tablespoons chopped fresh dill weed
2 tablespoons finely chopped onion
1 cup finely chopped raw cabbage
1 teaspoon minced garlic
2 teaspoons salt
½ cup **lemon** juice
1 tablespoon brown sugar
½ cup sour cream

Wash beets, put them in a large kettle of water with the salt, and bring to a boil. Boil, covered, for 45 minutes to 1 hour. Remove when tender, plunge into cold water, and rub off skins. Chop into ½-inch pieces. In a large saucepan, combine beets, broth, dill, onion, cabbage, garlic, salt, lemon juice, and brown sugar. Bring to a boil and cook over moderate heat, covered, for 25 minutes. Put in food processor or blender until smooth. Pour into soup bowls and garnish with a dollop of sour cream on each.

JAPANESE CLAM SOUP

Serves 6

1 cup water
1 teaspoon **lemon** juice
Small bunch watercress sprigs
6 cups water
1 small block of kelp (available at gourmet and Japanese vegetable stores)
1 dozen fresh cherrystone clams, cleaned and scrubbed
½ pound small mushrooms
2 tablespoons soy sauce
1 teaspoon salt
6 strips **lemon** zest (1 inch by ⅛ inch)
¼ onion, cut into very thin slices

Combine 1 cup water, lemon juice, and watercress in a small saucepan. Bring to a full boil and remove from heat. Place the sprigs on paper towels to drain; discard the water. Pour additional water, the kelp, and the clams in their shells into a large saucepan. Bring to a boil and remove and discard the kelp. Continue to boil the soup until the clams open, being sure to skim off any surface scum as it rises. Immediately add the mushrooms, soy sauce, and salt, and stir for 10 seconds. Remove from heat. To serve, pour the soup and two clams into each bowl. Float a strip of lemon zest and an onion slice in each bowl.

FRESH FRUIT SOUP

Serves 6

½ medium cantaloupe, seeded
2 cups fresh strawberries
¼ pound seedless grapes
¼ pound cherries, pitted
2 medium cooking apples, peeled, cored, and coarsely chopped
⅓ cup **lemon** juice
¼ cup sugar
3 cups water
¾ cup orange juice
½ cup sour cream

Remove the meat from the cantaloupe and cut into ½-inch bits. Wash the strawberries and grapes, picking off the stems and leaves.

Using a large saucepan, combine the cantaloupe, strawberries, grapes, cherries, chopped apple, lemon juice, sugar, and water. Bring to a boil. Lower heat and simmer, uncovered, for 10 minutes. Remove from heat and stir in the orange juice. Chill for 4 or 5 hours. Just before serving, add a dollop of sour cream to each bowl.

FISH

FISH

Fish and lemon are a partnership as familiar as pizza and pepperoni—and can be about as inspiring. A plate of freshly grilled trout can always be satisfied by a wedge of lemon; the puzzle is to present this match in some novel ways.

Spicy Swordfish does have a lot of spice, so serve it with a neutral backdrop, like rice. The same is true for Seafood Gumbo, although you can control the hotness by adding more or less Tabasco.

Lemony Flounder is more clever than it appears: as the flounder bakes, the mayonnaise melts into the lemon juice, seasonings, and fish juices, producing an aromatic sauce.

All fish dishes are fine warm-weather dishes, but we would particularly recommend Poached Salmon. Serve this dish with a good cold mayonnaise sauce, perhaps a salad with cucumbers or mushrooms, and open a bottle of dry white wine like a French Mâcon or a California Pinot Chardonnay. Voilà, a three-star meal.

BLUEFISH WITH PICKLED LEMONS

Serves 4

2 small onions, minced
2 tablespoons olive oil
4 bluefish fillets (or you can use sea trout, mackerel,
  or any other full-flavored fish)
4 medium tomatoes, peeled, seeded, and chopped
½ cup white wine
½ cup clam juice or fish stock
½ teaspoon dried thyme
2 pickled **lemons,** quartered (see page 77)
Freshly ground black pepper
Salt (optional)

Preheat oven to 375 degrees. Wilt onions until transparent in olive oil in a large flat roasting or lasagna pan big enough to hold fish in one layer. Turn off heat, lay in fish fillets, add tomatoes, wine, clam juice or fish stock, and thyme. Cover pan with foil and bake for about 20 minutes or until just before fish flakes moistly when touched with a fork. Add lemons and return to oven for 5 minutes. Grind pepper over, and add salt only if needed (clam juice and lemons are salty). Serve with couscous and cucumber and yogurt salad.

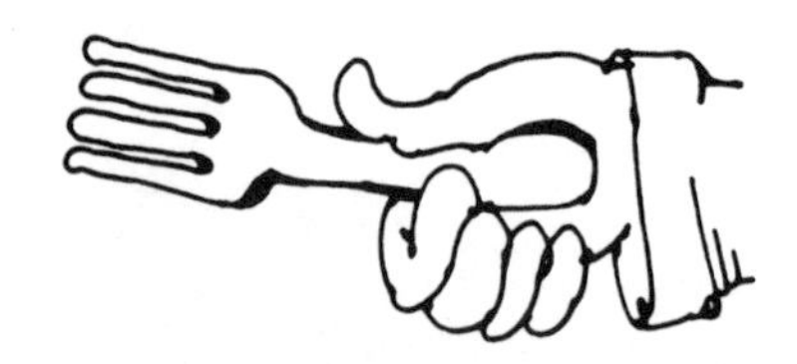

SPICY SWORDFISH

Serves 6

3 tablespoons **lemon** juice
3 tablespoons orange juice
$1/4$ cup soy sauce
3 tablespoons tomato paste
$1\frac{1}{2}$ teaspoons minced garlic
1 teaspoon dried oregano
1 teaspoon chopped fresh basil
3 teaspoons chopped fresh parsley
$1/2$ teaspoon salt
Freshly ground black pepper to taste
$1/3$ cup water
2-pound piece swordfish
1 **lemon** cut into wedges

Mix all ingredients except swordfish and whole lemon in a large baking dish. Add the piece of swordfish and baste so that the fish is covered with the marinade. Let marinate for 3 hours, turning occasionally. Place a piece of foil on broiler pan, and, when broiler is heated, place fish on foil and brush with marinade. Broil for 10 minutes, or until cooked through. Remove and pour pan juices over fish. Serve swordfish with lemon wedges.

POACHED SALMON

Serves 8

5 quarts water
1 pound fish trimmings
⅓ cup **lemon** juice
1 bay leaf
1 teaspoon whole black peppercorns
1 tablespoon salt
1 tablespoon butter
6-pound salmon, cleaned
8 **lemon** wedges

Fill a large saucepan with the water, add fish trimmings, lemon juice, bay leaf, peppercorns, and salt, and bring to a boil. Boil for 30 minutes, then pour liquid through a strainer into a large (12-quart) fish poacher. Discard trimmings, peppercorns, and bay leaf. Grease the tray of the poacher with the butter. Wash salmon under cold water and place in tray of poacher. Simmer the salmon for 30 minutes, making sure it is always covered by water. Remove fish from water and, while it is still warm, skin the top surface with a sharp knife. Carefully flip fish over and remove skin from the upturned side. Serve hot with lemon wedges and herbed butter, or cold with lemon wedges and herbed mayonnaise.

SALMON SOUFFLÉ

Serves 6

3 tablespoons butter
$\frac{1}{4}$ cup finely chopped onions
3 tablespoons flour
1 cup milk
5 egg yolks
$1\frac{1}{2}$ tablespoons tomato paste
Two 7-ounce cans salmon, drained
2 tablespoons finely chopped fresh dill weed
$\frac{1}{2}$ teaspoon finely chopped fresh basil
2 tablespoons **lemon** juice
$1\frac{1}{2}$ teaspoons salt
$\frac{1}{8}$ teaspoon cayenne pepper
6 egg whites
$\frac{1}{2}$ teaspoon salt
Freshly ground white pepper to taste

Preheat oven to 400 degrees. With 1 tablespoon of the butter, coat the bottom and sides of a 7- or 9-inch soufflé dish. In a saucepan, heat remaining butter and add onions. Simmer for 3 minutes. Mix in flour and blend well. Whisk in the milk and remove from heat. Beat the egg yolks into the mixture one at a time, and then stir in the tomato paste, salmon, dill, basil, lemon juice, salt, and cayenne. Let cool. Meanwhile, beat egg whites in a copper bowl until firmly peaked. Fold in salmon mixture, blending gently with a spatula while adding salt and pepper. Pour mixture into soufflé dish. Place in oven and immediately reduce heat to 375 degrees. Bake for 35 minutes and remove when soufflé rises into a light-brown puff. Serve at once.

SCROD IN LEMON BUTTER

Serves 4

6 tablespoons butter, melted
3 tablespoons **lemon** juice
2 pounds skinless fresh scrod fillets
½ teaspoon salt
Freshly ground black pepper to taste
3 tablespoons bread crumbs
1 **lemon,** cut into 4 wedges

Preheat broiler. Mix the butter and lemon juice in a large baking dish. Dip the fillets in this until they are well coated, and place side by side in baking dish. Sprinkle with salt and pepper. Put under broiler for 4 minutes, basting twice. Sprinkle bread crumbs on top of fillets and broil for 5 more minutes. Serve with lemon wedges.

LEMONY FLOUNDER

Serves 4

4 flounder fillets
¼ cup mayonnaise
½ cup **lemon** juice
Salt and freshly ground black pepper
1 **lemon,** cut into 12 thin slices
1 **lemon,** cut into 4 wedges

Preheat oven to 350 degrees. Wash fillets and pat dry. Smear ½ tablespoon of mayonnaise on each side of fish. Place each in generous-size piece of aluminum foil. Pour some lemon juice over each, and sprinkle with salt and pepper to taste. Top each with 3 slices of lemon. Wrap foil around each fish to form an individual packet. Cook for 15 to 20 minutes in oven. Serve with lemon wedges.

SEAFOOD GUMBO

Serves 8

½ cup butter
1 pound fresh okra, sliced into ¼-inch pieces
½ cup finely chopped green pepper
½ cup finely chopped onions
1 garlic clove, minced
3 tablespoons flour
5 cups chicken stock
1 pound canned Italian plum tomatoes, drained and chopped
6 sprigs parsley
1 teaspoon dried thyme
½ teaspoon dried tarragon
2 teaspoons salt
Freshly ground black pepper to taste
1 pound small raw shrimp, shelled and deveined
16 oysters, shucked
½ pound crabmeat
3 tablespoons **lemon** juice
2 teaspoons Worcestershire sauce
1 teaspoon hot sauce (Tabasco)
6 cups hot cooked rice

Heat ¼ cup of the butter in a large heavy frying pan. Add the okra, stir a few minutes over low heat, and set aside. Heat the remaining butter in a large heavy casserole. Slowly stir in the green pepper, onions, and garlic, and cook over low heat for 5 minutes. Slowly mix in the flour. Whisk the chicken stock in and add the okra, tomatoes, parsley, thyme, tarragon, salt, and pepper. Bring to a boil and cook, partially covered, over low heat for 35 minutes. Remove cover, add the shrimp, and simmer for 7 minutes. Add the oysters and crabmeat and continue to simmer for 3 minutes. Remove from heat. Add the lemon juice, Worcestershire sauce, and hot sauce and stir once more. To serve, ladle into soup bowls filled with a portion of cooked rice.

MUSSELS WITH TOMATO AND LEMON

Serves 4

2 quarts mussels
2 tablespoons butter
1 cup finely chopped onions
2 cups finely chopped tomatoes
2 teaspoons **lemon** juice
1/4 cup chopped fresh parsley
2 lemons, quartered

Scrub the mussels under cold water and remove the beards. Melt the butter in a heavy frying pan. Add the onions and cook over low heat for 4 to 6 minutes or until golden brown. Stir in the tomatoes and lemon juice. Place the mussels in the pan and turn the heat up to high. Stir and shake pan until all mussels open. Ladle into deep bowls and sprinkle the parsley over each bowl. Pass lemon wedges and serve with a loaf of hot crusty French bread.

POULTRY

POULTRY

Lemon and poultry have an affinity much like the more familiar fish and lemon; Lemon Game Hen is a good example. This recipe, which is easy enough to make as a cure for the Monday night blahs and elegant enough for a Saturday night dinner party, depends on the deceptively simple premise that lemon and game hen harmonize beyond reasonable expectations. They do. The hens become totally perfumed with lemon, creating a taste with surprising depth and richness.

Jane's Fried Chicken Slices with Lemon Sauce belongs to Jane Price—not because she invented it, but because she translated it from the Chinese for its English debut in this book. The recipe is quite unlike the lemon chicken served in most Chinese restaurants; we think it's much superior.

The Tandoori Murgh is, of course, Indian. Although it isn't orthodox unless the chicken is roasted in a tandoor, using a regular oven makes it easier to protect the chicken from terminal dehydration.

In Chicken with Lemony Mustard be sure to use coarse salt—because it gives the chicken skin a unique texture.

LEMON GAME HEN

Serves 6

6 Cornish game hens, fresh or frozen and defrosted
2 tablespoons coarse salt
6 garlic cloves, peeled
6 **lemons**
½ cup butter, melted

Preheat oven to 375 degrees. Rinse the hens under cold running water and dry. Rub the cavity of each bird with 1 teaspoon salt. Then place a garlic clove in each bird. Prick each lemon all over, ten times, with a sharp skewer. Place a pricked lemon inside each hen and arrange on a roasting pan.

Put the birds in the oven. Roast 45 minutes to 1 hour, basting every 10 minutes with butter, until birds are golden and tender. Remove the lemons and serve birds on a warmed platter.

SQUAB ROASTED WITH LEMON-SOY BUTTER

Serves 6

6 garlic cloves, peeled
6 squabs, rinsed and dried
1½ cups melted butter
3 tablespoons soy sauce
1½ tablespoons lemon juice
Freshly ground black pepper
Small bunch watercress

Preheat oven to 450 degrees. Place a garlic clove in the cavity of each bird and place in a roasting pan. Mix the butter, soy sauce, and lemon juice. Using a pastry brush, coat each bird with the butter mixture.

Put the birds in the oven and immediately turn down heat to 350 degrees. Baste every 5 minutes or so. Roast until birds are tender and golden, about 30 minutes.

Remove birds to a warmed platter, grind pepper over all, scatter watercress over all, and serve.

CHICKEN EMILIANA

Serves 6

3 pounds chicken breasts and thighs
3 cups orange juice
½ cup honey
3 **lemons**
1 small can green chilies, chopped (not hot chilies)
2 to 3 jalapeño chili peppers (most easily found bottled and pickled), seeded and chopped
12 preserved kumquats

Preheat oven to 375 degrees. Place the chicken skin side down in a roasting pan. In a large bowl, mix together the orange juice, honey, and juice of the lemons. Add the green chilies and the chili peppers. Pour mixture over chicken. Bake for 30 minutes. Turn chicken over and add kumquats. Bake another 20 minutes, basting every 5 minutes. Chicken is done when you can pierce it easily with a fork. Remove chicken and kumquats to a heated serving platter, spoon some of the sauce over all, and serve the remainder in a sauceboat.

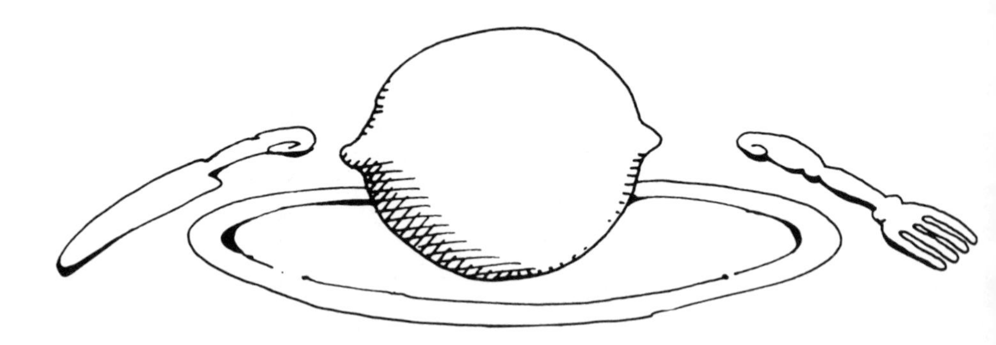

CHICKEN WITH LEMONY MUSTARD

Serves 6

3 pounds chicken breasts and thighs
$1\frac{1}{2}$ **lemons**
$\frac{1}{4}$ cup olive oil
$\frac{1}{4}$ cup Dijon-style mustard
Coarse salt (sea salt or kosher salt)
Scallions, cut lengthwise into slivers

Preheat broiler to moderately hot, about 400 degrees. Place the chicken skin side down in a broiler pan. Using a quarter of a lemon per chicken piece, drench pieces in lemon juice, squeezing and rubbing the lemon into the chicken. Gradually whisk olive oil into mustard, then coat the chicken tops with half the mixture, reserving the other half. Salt each piece with coarse salt. Broil for 10 to 15 minutes, until brown.

Turn pieces over and repeat the mustard mixture and salt. Broil another 10 minutes. Remove from broiler and place on a warmed serving platter. Sprinkle with scallions. Let the chicken rest a minute or two while you scrape the pan juices into a sauceboat.

TANDOORI MURGH

Serves 6 to 8

1 cup plain yogurt
5 garlic cloves, coarsely chopped
1 medium onion, coarsely chopped
$2/3$ cup **lemon** juice
$1\frac{1}{2}$ tablespoons grated fresh ginger
1 tablespoon ground coriander
1 tablespoon ground cumin
1 tablespoon cayenne pepper
1 teaspoon salt
2 chickens, $2\frac{1}{2}$ to 3 pounds each, quartered

Mix together the yogurt, garlic, onion, lemon juice, ginger, coriander, cumin, cayenne, and salt. Using a sharp knife, cut a slit in each chicken piece $1/2$ inch deep and 1 inch long. Place the chicken in a casserole dish and pour in yogurt mixture. Cover and marinate in the refrigerator for 48 hours. Preheat oven to 375 degrees. When you are ready to roast the chicken, put pieces in a shallow roasting pan in a single layer. Cook, basting the pieces with the yogurt mixture every 10 minutes or so, until chicken is tender and golden—about 50 minutes.

JANE'S FRIED CHICKEN SLICES WITH LEMON SAUCE

Serves 4

1 pound chicken breasts, skinned and boned

MARINADE

½ tablespoon dry sherry
1 tablespoon soy sauce
1 tablespoon cornstarch
3 tablespoons cold water
1 egg yolk
½ teaspoon black pepper

6 tablespoons cornstarch
3 tablespoons flour
1¼ cups peanut oil

LEMON SAUCE

3 tablespoons sugar
¼ cup **lemon** juice
3 tablespoons chicken bouillon
2 teaspoons cornstarch
1 teaspoon sesame oil
½ teaspoon salt

1 teaspoon peanut oil
1 **lemon,** cut into paper-thin slices

Cut chicken breasts into strips 1½ inches wide, 2 inches long. Combine the ingredients for the marinade in a small bowl..

Marinate the chicken strips for 10 minutes. Mix the 6 tablespoons cornstarch and 3 tablespoons flour on a plate. Coat each strip of chicken. In a wok or frying pan, heat the peanut oil over low heat and then deep-fry the strips a few at a time for about 30 seconds, or until golden. Drain. Increase the temperature of the oil until quite hot, and deep-fry the strips again for 10 seconds. Drain and remove to heated platter.

Mix the ingredients for lemon sauce in a small bowl. In a wok or small frying pan, heat 1 teaspoon of peanut oil and pour in sauce. Stir-fry until sauce boils and thickens, about 3 minutes.

Pour the sauce over the fried chicken strips and serve hot, surrounded by lemon slices.

WHITE BREASTS OF CHICKEN

Serves 6

½ cup butter, melted
1 teaspoon salt
1 teaspoon white pepper
6 whole chicken breasts, skinned, boned, and split in half
1 lemon
¼ cup coarsely chopped fresh parsley or watercress

Preheat oven to 450 degrees. Combine the melted butter, salt, and pepper in a small, deep bowl. Dip each chicken breast in the bowl so that it is thoroughly coated with butter. Arrange the breasts in a shallow baking dish in a single layer. Cut the lemon in quarters and squeeze the juice over the breasts. Bake in oven for 10 minutes, until breasts are tender but not brown, so that they appear to have been poached instead of baked. Remove from oven and sprinkle the parsley or watercress over all.

CHICKEN LEMON KIEV

Serves 6

1 cup butter
2 tablespoons **lemon** juice
4 scallions, finely chopped
1 tablespoon chopped chives
1 teaspoon dried tarragon, crushed
¼ teaspoon finely chopped **lemon** zest
1 teaspoon salt
½ teaspoon pepper
6 whole chicken breasts, skinned, boned, and split in half
¾ cup flour
Dash of salt and black pepper
3 eggs
1 teaspoon **lemon** juice
1 teaspoon peanut oil
8 ounces unseasoned bread crumbs
1 quart peanut oil
Watercress or parsley

Bring butter to room temperature, then blend in 2 tablespoons lemon juice, scallions, chives, tarragon, lemon zest, 1 teaspoon salt, and ¼ teaspoon pepper. Place the mixture on a piece of aluminum foil and shape into a square about ¾ inches thick. Place in freezer for 1 hour.

Meanwhile, wash and dry the chicken pieces and pull or cut

open the small pocket near the split. Place chicken pieces between two pieces of waxed paper and flatten with a mallet until ¼ to ½ inch thick.

Take the butter out of the freezer, slice into 4-inch-by-1-inch pieces, and place in the center of each chicken piece. Fold each flap over, making sure that the butter is covered completely: place the pieces in the palm on one hand, and squeeze with the other hand until the chicken remains sealed and closed. Handling gently, dredge chicken pieces in the flour mixed with dash of salt and pepper. Mix the eggs, 1 teaspoon lemon juice, and 1 teaspoon peanut oil into a bowl, and dip the dredged chicken into this mixture. Roll the chicken in bread crumbs to coat well.

Pour the quart of peanut oil into a large heavy skillet. Heat oil and put the chicken in the skillet. Keep the oil very hot while the chicken cooks, and gently turn over after 3½ minutes. Cook for another 3½ minutes and remove to a warmed platter lined with paper towels. Remove towels before serving, and garnish the chicken with watercress or parsley.

ROAST DUCK WITH SAUSAGE

Serves 4

2 lemons
5-pound duck
1 large garlic clove
1 cup uncooked long-grain rice
½ pound garlicky pork sausages
2 tablespoons vegetable oil
2 onions, minced
2 tablespoons **lemon** juice
¼ cup minced fresh parsley

Preheat oven to 450 degrees. Cut the zest off one of the lemons; save both the lemon and the zest.

Wash the duck under cold running water, then dry thoroughly. Place the duck on a rack in a roasting pan. Pop the garlic and the lemon zest into the cavity, and squeeze the peeled lemon all over the duck. Place in the oven and immediately turn down the heat to 350 degrees. Roast until brown and tender, about 1½ hours.

As the duck is roasting, prepare the rice and the sausages. Bring 2½ cups of water to a boil in a heavy saucepan. Drop the rice in, cover, and turn the heat down to low. Let cook for 20 minutes until the grains are tender but still firm to the bite. Set aside.

Prick the sausages in 3 or 4 places with a fork and put into a large frying pan. Add enough cold water to cover the sausages, and bring to a boil. Reduce heat to a simmer for 5 minutes. Drain the sausages and cut into thin slices.

Pour the oil into a large heavy frying pan and heat over high heat. Add the sausages and cook for 4 minutes, turning the slices often. Turn heat down to low, add the onions, and cook another 4 minutes, until the onions are mushy but not brown. Then add the rice and lemon juice. Stir until combined. Taste for seasoning.

Make a bed of rice on a large heated serving platter. Carve the duck into serving pieces and arrange, skin side up, on the rice. Scatter the parsley over all, and decorate the edge with slices from the remaining lemon.

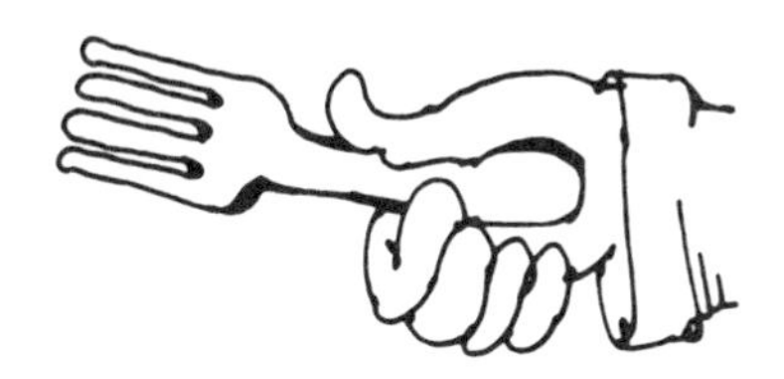

DUCK BIGARADE

Serves 4

5-pound duck
2 tablespoons sugar
¼ cup vinegar
1 cup veal or chicken stock
1 tablespoon **lemon** juice
1 cup orange juice
2 tablespoons grated orange zest
Salt and black pepper to taste
1 orange, halved and cut into paper-thin slices

Preheat oven to 450 degrees. Wash the duck under cold running water and dry thoroughly. Put the bird on a rack in a roasting pan, place it in the oven, and immediately turn down the heat to 350 degrees. Roast until brown and tender, about 1½ hours. Skim off the fat and throw it away, but save the pan drippings.

Melt the sugar in the vinegar in a heavy saucepan over low heat, stirring constantly so that the sugar doesn't burn. Add the pan drippings and cook for 5 minutes, stirring frequently. Then add the stock, lemon juice, orange juice, and orange zest. Simmer another 5 minutes, then add salt and pepper.

Serve the duck on a warmed platter decorated with orange slices. Pass the sauce separately in a sauceboat.

MEAT

MEAT

We present two recipes for veal scaloppine with lemon because both approaches claim to be exemplar and both taste exemplary. The chief difference is in choice of wine to flatter the veal. Kevin's Veal is a little less lemony, so you might pour a wine on the mellow side, like Soave. And Frascati would be a great match for Veal Scaloppine with Lemon Sauce.

Leg of Lamb Niçoise is a Provençal dish with a big, hearty flavor. Since the sauce is a major component, it is best to have an efficient blotter, like rice or noodles, with the lamb.

Tangy Lamb Curry Custard is a bit different from traditional Indian curries: it is baked and bound together in a custard rather than with yogurt, and it contains fruit and nuts as well as meat. Tangy Lamb Curry Custard needs nothing more than some hot lemon pickle alongside and a cool cucumber salad to follow.

German-Style Ham is also an unusual recipe. The marinade of juniper berries, lemon, and spices creates a baked ham with a slightly gamy, rich taste.

KEVIN'S VEAL

Serves 6

½ cup butter
1¼ pounds veal scaloppine, sliced very thin and pounded flat
1 cup flour
Salt to taste
¾ cup white wine
1½ tablespoons **lemon** juice
Parsley sprigs

Heat the butter in a large heavy frying pan until hot. While butter is heating, dip the veal in the flour, lightly coating both sides. Put the veal in the frying pan, being careful not to overlap any slices. Cook the veal until lightly browned on one side, then flip slices over and lightly brown on the other side. If your slices are very thin, figure 30 seconds for each side. Remove the veal to a warm serving dish and add salt.

Add the wine to the frying pan and let it cook down a few minutes. Add the lemon juice and stir. Cook for 1 minute. Pour the sauce over the veal and add parsley sprigs. Serve at once.

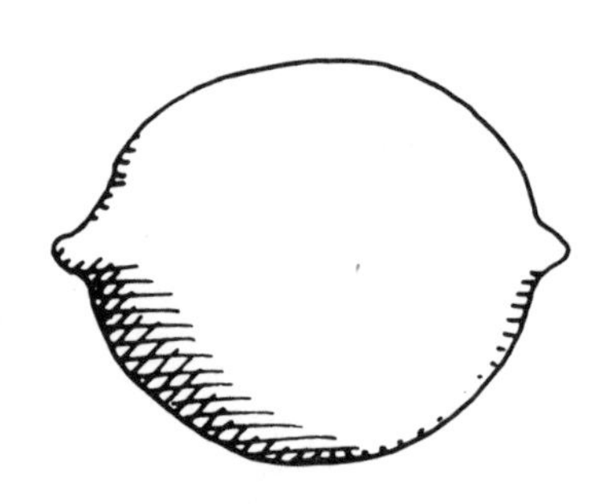

VEAL SCALOPPINE WITH LEMON SAUCE

Serves 6

2 tablespoons vegetable oil
⅓ cup butter
1¼ pounds veal scaloppine, sliced very thin and pounded flat
1 cup flour
Salt and black pepper to taste
3 tablespoons **lemon** juice
3 tablespoons chopped fresh parsley
1 lemon, sliced very thin

Heat the oil and half the butter in a large heavy frying pan until sizzling hot. While the butter and oil are heating, dip the veal in the flour, lightly coating both sides. Put the veal in the frying pan, being careful not to overlap any slices. Cook the veal until lightly browned on one side, then flip slices over and lightly brown on the other side. If your slices are very thin, figure on 30 seconds for each side. Remove the veal to a warm serving dish and add salt and pepper.

Lower the heat and add the lemon juice, stirring and scraping up the cooking remnants. Add the remaining butter, stir until melted, sprinkle in the parsley, and stir briefly. Pour the sauce over the veal and decorate with the lemon slices. Serve immediately.

WIENER SCHNITZEL

Serves 6

2½ pounds veal leg, sliced into cutlets ¼ inch thick
1¼ cups **lemon** juice (approximately 7 to 8 lemons)
Salt and black pepper to taste
3 eggs
3 tablespoons water
½ cup flour
1½ cups fine bread crumbs
¼ cup vegetable oil
1 **lemon,** cut into 6 wedges

Marinate the veal in lemon juice for 30 minutes. Remove the cutlets from the lemon juice and dry thoroughly. Salt and pepper them liberally. Whisk the eggs with the water. Dip cutlets into the mixture, then dip them into the flour, then into the bread crumbs, coating both sides. Shake off any excess crumbs and refrigerate for 30 minutes.

Heat the oil in a large heavy frying pan over medium heat. When oil is hot, add the cutlets. Brown on each side, about 4 minutes, then remove to a heated platter and decorate with lemon wedges. Serve immediately.

OSSO BUCO

Serves 6 to 8

6 to 7 pounds veal shanks, sawed into eight 2-inch pieces
¾ cup flour
¼ cup butter
1 cup minced onions
⅔ cup scraped and minced carrots
½ cup minced celery
1 garlic clove, minced
2 strips **lemon** zest
¼ cup olive oil
1 cup dry white wine
1 cup beef bouillon
½ teaspoon dried thyme
1½ pounds canned Italian tomatoes, coarsely chopped
    in their juice
2 bay leaves
Freshly ground black pepper to taste

GREMOLATA

1 medium garlic clove, minced
1 tablespoon grated **lemon** zest
¼ cup minced fresh parsley

Preheat oven to 350 degrees. Dip the veal shanks in the flour, coating all sides. Using a large, heavy, shallow casserole dish which you can cover tightly, heat the butter. When hot, cook the onions, carrots, celery, and garlic over moderate heat until they have softened, about 10 minutes. Add the lemon zest and remove from heat.

Heat the olive oil in a large heavy frying pan. When hot, add the pieces of floured veal and brown on all sides. Remove the veal to the casserole dish and arrange on the bed of vegetables in one layer, standing pieces upright so that the marrow will not fall out during cooking.

Pour off all but a teaspoon or two of oil from the frying pan. Add the wine and bring to a boil, scraping the bottom and sides of the pan. Boil for 3 to 4 minutes, then add the beef bouillon and thyme. Bring to a boil while stirring constantly. Pour over the veal in the casserole. Scatter the chopped tomatoes and their juice, bay leaves, and pepper over the veal. The braising liquid should stand at least halfway up the sides of the veal.

Bring the casserole to a boil on top of the stove, then cover and place in the oven. Stock should be maintained at a very slight simmer. Braise for 1½ hours, basting every 15 minutes. Meanwhile, combine and mix lightly gremolata ingredients.

When ready to serve, carefully remove the veal to a heated platter. Strain the vegetables from the sauce and boil it down by half. Pour the reduced sauce over the veal and sprinkle gremolata over all.

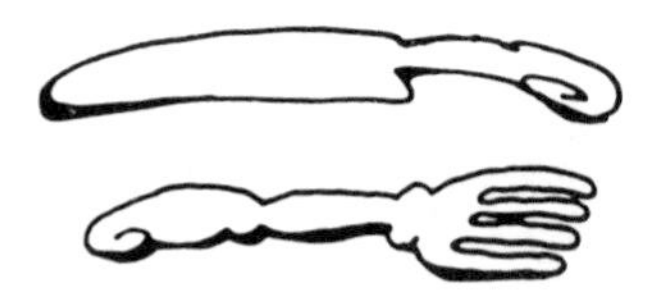

TANGY LAMB CURRY CUSTARD

Serves 6

2 slices white bread, crusts removed
2 cups milk
¼ cup butter
2 pounds ground lamb
2 cups coarsely chopped onions
1 garlic clove, minced
2 tablespoons curry powder
1 teaspoon turmeric
½ teaspoon fenugreek
1 tablespoon light-brown sugar
1 teaspoon salt
½ teaspoon cayenne pepper
⅓ cup **lemon** juice
4 eggs
½ cup raisins
½ cup blanched almonds, coarsely chopped

Preheat oven to 300 degrees. Shred the bread and put in a bowl with the milk to soak. Melt half of the butter over moderate heat in a large frying pan, then add the lamb. Mash the lamb with the back of a fork until it is cooked and separated into small bits. Transfer the lamb with a slotted spoon to a large bowl. Discard all the fat. Melt the remaining butter in frying pan and add the onions and garlic. Cook over moderate heat until onions are soft and barely beige—about 6 minutes. Then add curry powder, turmeric, fenugreek, brown sugar, salt, and cayenne. Stir over heat for 3 minutes. Pour in the lemon juice, bring to a boil, stir a couple of seconds, then pour the mixture into the bowl with the lamb, and mix.

Drain the bread, reserving the milk, and squeeze the bread until dry. Add the bread, 1 egg, raisins, and almonds to the lamb. Beat the mixture until everything is completely combined. Taste for seasoning. Put the lamb mixture, loosely packed, into a deep baking dish.

Using a whisk, beat together the remaining 3 eggs and the reserved milk. Pour into the lamb mixture. Bake in the middle of the oven for 35 to 40 minutes or until the custard has set.

LEG OF LAMB NIÇOISE

Serves 6 to 8

6-pound leg of lamb
5 large garlic cloves, peeled
2 tablespoons olive oil
1 celery stalk, minced
1 carrot, scraped and minced
1 cup minced onions
1 cup beef bouillon
Bouquet garni (thyme, basil, bay leaf, parsley)
1 cup chicken bouillon
1 cup black olives, Mediterranean-style, pitted and chopped
2 tablespoons drained and mashed anchovy fillets
2 tablespoons minced fresh parsley
Juice of ½ medium **lemon**
Salt and freshly ground black pepper to taste

Preheat oven to 400 degrees. Using a sharp knife, make about six incisions near and along the bone of the lamb. Cut 2 garlic cloves into slivers and insert them into incisions. Mince the remaining 3 garlic cloves, and reserve. Heat the oil in a large heavy kettle or iron casserole. Add the garlic, celery, carrot, and onions. Cook slowly for 10 minutes or until vegetables are soft but not brown. Add ½ cup of beef bouillon and bouquet garni, heat through, and set aside. Insert meat thermometer in lamb and put lamb in a roasting pan. Set in oven to brown for 15 minutes. Remove the pan from oven and turn down heat to 375 degrees. Put the lamb in the kettle or casserole, on top of the vegetables; cover, and set in the oven. Cook for 1 hour and 25 minutes, or until meat thermometer reads 145 degrees.

Remove lamb to a warmed serving platter. Strain the cooking liquid into a bowl, pressing to extract all the liquid before discarding vegetables. Skim off as much fat as possible. Return the cooking liquid to the kettle. Add the remaining beef bouillon and the chicken bouillon. Bring to a boil. Reduce heat slightly, and add olives, anchovies, parsley, and lemon juice. Keep stirring until thoroughly heated. Taste for seasoning.

Carve the lamb, and pass the sauce separately in a warmed sauceboat.

SHISH KEBAB

Serves 6

MARINADE

1 cup olive oil
⅓ cup **lemon** juice
½ cup dry red wine
2 garlic cloves, crushed
2 large bay leaves
1 teaspoon salt
½ teaspoon black pepper

3 pounds boneless lamb, cut into 1-inch cubes
18 cherry tomatoes
18 pearl onions, peeled
18 large mushroom caps

Combine the ingredients for the marinade in a large bowl. Add the lamb chunks, and mix well. Cover and refrigerate for 12 to 24 hours.

Preheat broiler. Using six skewers, each at least 12 inches long, skewer a chunk of lamb, a tomato, an onion, a mushroom cap; then repeat. When broiler is hot, place the skewers on a broiling pan about 3 inches from the heat. Broil, turning often, until meat is done, about 15 to 20 minutes. Serve over rice or serve unadorned, but serve immediately.

LAMB KORMA CURRY

Serves 6

2 pounds boneless leg of lamb
1/4 cup vegetable oil
2 medium onions, chopped
1 garlic clove, crushed
2 tablespoons ground coriander
1 teaspoon cumin
1 teaspoon turmeric
1-inch piece fresh ginger, grated
1 teaspoon cayenne pepper
6 cloves
1/2 teaspoon ground cardamom
1 cup plain yogurt
1 cup water
Two 1-inch cinnamon sticks
2 teaspoons salt
1 teaspoon **lemon** juice
3 tablespoons grated coconut

Trim the lamb and cut into 7-inch chunks. Heat the oil in a large skillet. Add the onions and garlic and cook, stirring, until the onions are browned and tender. Scrape the onions and garlic to one side of the skillet, add the lamb chunks, and brown on all sides. Remove from heat, add the coriander, cumin, turmeric, ginger, cayenne, cloves, and cardamom. Mix thoroughly, turn heat down to moderate, return skillet to burner and cook for a minute, stirring constantly. Add the yogurt, water, cinnamon, and salt. Mix well. Cover and simmer slowly until lamb is tender, about an hour, adding more water if necessary.

Just before serving, stir in the lemon juice and coconut. Lamb Korma Curry is transcendental when spooned over Lemon Rice (page 95).

SAUTÉED KIDNEYS

Serves 6

10 lamb kidneys
3 tablespoons butter
¼ cup oil
3 garlic cloves, minced
3 tablespoons minced fresh parsley
3 tablespoons **lemon** juice
Salt and freshly ground black pepper

Remove membranes from kidneys. Cut the kidneys crosswise into very thin slices. In a large heavy frying pan, melt the butter and oil together over moderately high heat. As soon as the foaming has stopped, add the kidneys. Cook for 2 minutes, turning the kidneys frequently. Then add the garlic and parsley, and cook for another 2 minutes, still stirring. Pour in the lemon juice, let it come to a boil, then take the frying pan off the heat. Season with salt and pepper to taste and serve at once.

GERMAN-STYLE HAM

Serves 8 to 10

½ cup wine vinegar
2 cups dry red wine
1 cup minced onions
2 tablespoons crushed juniper berries
2 tablespoons grated **lemon** zest
3 bay leaves
1 teaspoon dried thyme
1 teaspoon whole cloves
½ teaspoon ground allspice
½ teaspoon ground ginger
1½ teaspoons black pepper
6-pound ham, uncooked, unsmoked, and trimmed of fat

Make the marinade by putting all the ingredients except the ham into a bowl and mixing them. Place the ham in a deep heavy casserole and pour the marinade over it. Cover and refrigerate for 24 hours; turn the ham over three or four times during marination.

Preheat the oven to 450 degrees. Remove the ham from the marinade and insert a meat thermometer. Place in a large roasting pan and put into the oven. Immediately turn the heat down to 350 degrees. Cook, uncovered, for 35 minutes to the pound or until the meat thermometer hits 180 degrees, basting every 30 minutes with the marinade.

SALAD
VINEGAR
S

SALAD AND SUCH

Why bother mixing oil, vinegar, and a packet of dried seasoning when a fresh lemon dressing is almost as fast and much more delicious? Lemon is not only an aesthetic substitute for vinegar, it's a necessary one when you're serving wine: salad dressing containing vinegar destroys the subtlety of even the most robust bottle. Although we think lemon is always a better bet for salad dressing, the salad recipes which follow use lemon as the major ingredient because vinegar would be unacceptably harsh.

The recipe for Shrimp with Lemon and Chili is a Thai dish (Pra Koong) which appears here in English for the first time thanks to Emmy Smith. It's very spicy and would be a fine main dish for a summer lunch or as part of a cold buffet.

In the Orange, Lemon, and Black Olive Salad, the fruit is an actual part of the salad, creating a knockout color combination. But unless you want your stomach knocked out by the acid, serve it with absorbent bread.

Both Sweet Lemon Pickle and Hot Lemon Pickle are excellent accompaniments to curries of any kind, but especially to beef or lamb. Likewise, Lois's Dried-Fruit Compote completes a traditional menu with any unadorned roast meat.

We've also included instructions for making Preserved Lemons. Moroccans add preserved lemons to stews and salads. We approve especially for lamb stews in which the pickled taste of one preserved lemon can lift the familiar dish into a more exotic category.

SALAD DRESSING

Makes 5 ounces

½ cup olive oil
3 tablespoons **lemon** juice
1 teaspoon prepared Dijon-style mustard
1 garlic clove, finely minced
Salt and freshly ground black pepper to taste

Combine all ingredients in a small bowl or jar and whisk or shake for 2 minutes. Refrigerate until ready to pour over salad greens. This can be kept in the refrigerator for a week—the flavor becomes more and more intense.

VARIATIONS:

Add ½ teaspoon chopped fresh basil and parsley
Add ½ teaspoon crushed dried tarragon

EMMY'S SHRIMP WITH LEMON AND CHILI

Serves 6

1 pound cleaned, shelled shrimp
1 tablespoon dark chili paste*
1 tablespoon fish sauce*
3 tablespoons **lemon** juice
1 small onion, sliced thin
2 scallions, chopped (white part only)
Cayenne pepper to taste
2 medium cucumbers, peeled and sliced thin
1 small head of lettuce, torn into salad-size pieces

Boil shrimp for 3 minutes in large kettle of water. Drain and place in a large bowl. Mix the following ingredients with the shrimp: chili paste, fish sauce, lemon juice, onion, scallions, cayenne, and cucumbers. Remove to a large salad bowl and toss with the lettuce.

* Available in any food specialty store

WHITE BEAN AND TUNA SALAD

Serves 6

One 20-ounce can white beans
One 6½-ounce can oil-packed tuna, drained
1 small Bermuda onion, sliced paper-thin
¼ cup olive oil
¼ cup **lemon** juice
Coarse salt and freshly ground black pepper to taste

Drain the beans and put in a salad bowl. Put the tuna in salad bowl, breaking it into little bits with a fork. Add the onion. Add the oil and toss thoroughly. Pour the lemon juice over all, and toss again. Right before serving, add salt and pepper and toss again.

GREEN BEAN SALAD

Serves 6

1½ pounds fresh green beans
1 tablespoon salt
¼ cup olive oil
Juice of 1 **lemon**
1 tablespoon coarse salt
Freshly ground black pepper to taste

Snap off the ends of the beans and pull off any strings. Soak them in cold water for 5 minutes, then drain. Bring 5 quarts of water to a rolling boil, add 1 tablespoon salt, and drop in the beans. Let beans cook at a moderate boil until they are tender but still crisp and firm—anywhere from 6 to 10 minutes, depending on how young the beans are. As soon as they are cooked, drain them.

Put the beans in a salad bowl and add the olive oil. Toss so that each bean is coated and shiny-looking. Add the lemon juice and toss again. Sprinkle on the coarse salt and pepper and serve.

MILWAUKEE POTATO SALAD

Serves 6

2½ pounds medium boiling potatoes
1 tablespoon dry vermouth
½ cup chopped scallions
⅔ cup chicken stock
¼ cup vegetable oil
2 teaspoons hot prepared mustard
2 teaspoons salt
1 teaspoon black pepper
2 tablespoons **lemon** juice

Peel the potatoes and drop them into enough salted, boiling water to cover. Boil for 20 to 30 minutes or long enough so that you can easily prick them with a fork. Drain. Slice the potatoes into ¼-inch slices. Sprinkle vermouth over them, and set aside in a covered bowl.

In a saucepan, combine the scallions, chicken stock, oil, mustard, salt, and pepper. Bring to a boil, stirring occasionally.

Reduce heat and simmer for 8 minutes. Remove the saucepan from heat and stir in the lemon juice.

Pour the sauce over the potatoes, turning them carefully with a rubber spatula so that every slice is coated. After the potatoes have cooled to room temperature, correct the seasoning.

ORANGE, LEMON, AND BLACK OLIVE SALAD

Serves 6

3 oranges, peeled and sectioned
1 **lemon,** peeled and sectioned
1¼ cups large black olives (Mediterranean-style), pitted
1 small Bermuda onion, sliced paper-thin

DRESSING

3 tablespoons olive oil
1 garlic clove, minced
¼ teaspoon sugar
⅛ teaspoon cumin
¼ cup chopped fresh parsley
Salt and pepper to taste

Arrange the orange and lemon sections with the olives and onion in a glass bowl. Make the dressing by putting all the remaining ingredients in a small mixing bowl and beating vigorously for a minute or two. When you're ready to serve, dress the salad and toss.

RAW BEEF AND TOMATO SALAD

Serves 6 to 8

2 pounds raw lean sirloin
2 garlic cloves, chopped
Juice of 4 lemons
½ cup olive oil
Salt and black pepper to taste
4 medium tomatoes, sliced thin
¼ cup coarsely chopped fresh basil

Cut the raw beef into paper-thin slices, about 1 inch long. Combine the garlic, lemon juice, oil, salt, and pepper, and pour over the beef in a deep bowl. Stir the beef to make sure every slice is well coated. Marinate in the refrigerator for 3 to 4 hours, stirring occasionally. When you're ready to serve, remove the beef with a slotted spoon and mound in the middle of a dish, surround with tomato slices, and sprinkle the basil over all.

SCALLOPS AND MUSHROOMS

Serves 6 to 8

1 pound fresh bay scallops
¼ cup minced scallions
Pinch of dried thyme
1 bay leaf
Salt and white pepper to taste
1 pound mushrooms
¼ cup **lemon** juice
½ cup olive oil
¼ teaspoon dried mustard
¼ cup chopped fresh parsley

Place the scallops, scallions, thyme, bay leaf, salt, and pepper in a saucepan. Add enough water to cover and bring to a simmer. Put the lid on the saucepan and simmer for 5 minutes. Using a slotted spoon, transfer the scallops to a bowl.

Thinly slice the mushrooms into a second bowl.

Put the lemon juice in a small bowl and add the olive oil in a thin stream while beating vigorously with a whisk. Add mustard, salt, and pepper. Dress the scallops with half of the dressing, the mushrooms with the other half. Chill for 2 hours. Just before serving, mix the scallops and mushrooms together in a big bowl, then sprinkle the parsley over all.

SWEET LEMON PICKLE

Makes about 1 quart

½ cup brown sugar
1 teaspoon cayenne pepper
½ teaspoon turmeric
⅛ teaspoon cumin
⅛ teaspoon fenugreek
2 tablespoons salt
¼ cup vegetable oil, heated and cooled
6 **lemons**, quartered lengthwise

Combine the brown sugar, cayenne, turmeric, cumin, fenugreek, and salt in a Mason-type jar. Pour in the oil and mix until blended. Add the lemons, cover tightly, and give the jar a good shake. Keep the jar in a warm place, shaking every so often, for 3 or 4 days. The lemon pickle is ready to eat when the lemon skins are tender enough to be cut by a fork.

HOT LEMON PICKLE

Makes about 1 quart

1 teaspoon mustard seed
1 teaspoon cumin
1 tablespoon cayenne pepper
1 tablespoon salt
2 cups vegetable oil
6 **lemons**, quartered

Mix the spices and salt together in a glass jar and set aside. Boil and then cool the oil. Pour the oil into the glass jar, shake, then add the lemons, squashing them in. Cover tightly and keep jar in a warm place for 4 to 7 days. Can be served chilled or at room temperature with curry dishes.

PRESERVED LEMONS

Makes about 1 quart

6 medium **lemons**
¼ to ½ cup salt
Fresh **lemon** juice (about 1 cup)

Quarter each lemon lengthwise from the top to within ¼ inch from the bottom so that lemons can be reshaped. Sprinkle the inside of the lemons with salt. Sprinkle a thin layer of salt on the bottom of a sterile quart-size jar, then pack the lemons in the jar—pushing them down upon each other so that their juices are squeezed out. Sprinkle in any remaining salt. Add enough fresh lemon juice to cover, but make sure you will have some air space at the top. Seal and give a good shake. Let stand in a warm place for at least 2 weeks, shaking every so often. Rinse lemons before using to remove briny taste.

MINT CHUTNEY

Makes about $1/2$ cup

2 cups fresh mint leaves
1 small white onion, minced
1-inch piece ginger root, chopped
1 teaspoon cayenne pepper
1 teaspoon salt
2 teaspoons sugar
Juice of 1 medium **lemon**

Using a food blender, or mortar and pestle, blend together mint leaves, onion, and ginger root. Then mix or blend in the cayenne, salt, sugar, and lemon juice. Refrigerate for at least 8 hours.

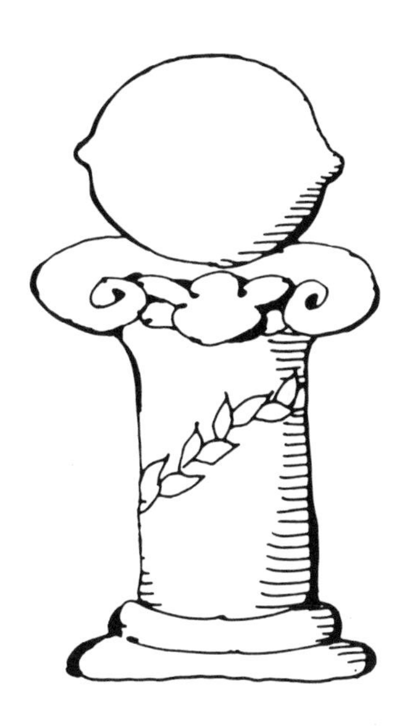

LOIS'S DRIED-FRUIT COMPOTE

Serves 10 to 12

1½ pounds dried fruit: apricots, prunes, currants, white raisins, and/or apples
2 cups sugar
2 medium cinnamon sticks
4 cloves
⅛ teaspoon allspice
Zest of 1 **lemon,** cut into long, narrow strips
Splash of dark rum

Soak the dried fruit overnight in a deep bowl with enough water to cover. Drain the fruit and adjust the water to make 1 quart. Pour the water into a big heavy saucepan and add the sugar, cinnamon, cloves, allspice, and lemon zest. Bring to a boil, stirring until sugar is dissolved. Add the fruit, reduce heat, and simmer for 15 minutes. Using a slotted spoon, remove the fruit to a serving dish and boil the syrup over high heat until it thickens a bit, 7 to 10 minutes, and then add the rum. Pour the syrup over the fruit. (Remember to remove the cinnamon sticks and the cloves.) Serve chilled. If compote is kept covered and chilled, it will keep for weeks.

VEGETABLES

VEGETABLES

One result of the recent turn toward health and simplicity in food has been an interest in plain vegetables. We welcome the disappearance of Vegetable-Medley-Smothered-Under-Brown-Sugar-Sauce. But the apparent replacement—great mounds of undercooked, underseasoned rutabaga—is hardly an improvement. We offer, then, a middle road: vegetables that are still recognizable, but are seasoned with some sophistication.

The first two recipes, Golden Potatoes and Broiled Tomatoes, are first-rate accompaniments to fairly uncomplicated dishes like roasts or broiled meat or fish; they should be served with the entrée.

Bernadette's Asparagus, Dilled Lima Beans, and even cold Mushrooms and Endives, however, have enough interest to be served as a separate vegetable course. These recipes would complement either simple or rich menus.

GOLDEN POTATOES

Serves 6

2 tablespoons butter
2 tablespoons olive oil
4 cups coarsely chopped potatoes
1 tablespoon finely chopped onion
2 tablespoons finely chopped fresh parsley
2 teaspoons **lemon** juice
½ teaspoon salt
Freshly ground black pepper to taste
⅓ cup light cream

In a large heavy frying pan mix the butter and olive oil over a low heat. Meanwhile, combine potatoes, onion, parsley, lemon juice, salt, and pepper into a large bowl. Place in frying pan and with spatula form 6 individual cakes. Slowly sauté the potatoes over low heat. When the potatoes are golden brown on the bottom, which should take 10 to 15 minutes, turn them over and immediately pour cream over them. Cook over moderate heat until golden brown, about 10 minutes, and serve.

BROILED TOMATOES

Serves 4

4 very ripe red tomatoes
4 teaspoons **lemon** juice
1 teaspoon finely chopped fresh basil
1 teaspoon finely chopped fresh tarragon
½ teaspoon finely chopped garlic
2 tablespoons butter, cut into bits
Salt and freshly ground black pepper to taste

Preheat oven to 375 degrees. Wash and slice the tomatoes in half, being sure to cut out the stem section. Pour ½ teaspoon of lemon juice over each inner half of tomato, and sprinkle a little basil, tarragon, garlic, butter, salt, and pepper on each. Place them on a greased baking sheet and cook 10 minutes or until soft. Turn oven to broil, and place tomatoes under broiler for 2 minutes or until golden on top. Serve immediately.

LEMON ARTICHOKES

Serves 4 to 6

⅓ cup olive oil
2 teaspoons salt
¼ teaspoon black pepper
1 **lemon,** cut in ½-inch slices
1 bay leaf
2 cloves garlic, halved
4 to 6 globe artichokes
½ **lemon**

SAUCE

2 tablespoons **lemon** juice
¾ cup butter, melted
2 tablespoons olive oil

Place ⅓ cup olive oil, salt, pepper, lemon slices, bay leaf, and garlic in a large pot. Add 4 quarts water and bring to a boil. Cut off the bottom stems of the artichokes and snip the ends of the leaves. Rub cut ends with the lemon half to prevent discoloration. Place artichokes under running water and drain. Drop in boiling water. Simmer uncovered for 45 minutes or until a bottom leaf may be easily detached. Remove artichokes, place upside down to drain, let cool, then chill in refrigerator.

Whisk lemon juice, melted butter, and olive oil in a bowl until well blended, and serve in small dishes with each artichoke.

CHARLIE'S BROCCOLI

Serves 8

1 teaspoon butter
1½ pounds fresh broccoli
3 tablespoons butter
3 tablespoons flour
½ pint heavy cream
1 cup beef stock
3 tablespoons dry white wine
¼ cup **lemon** juice
1 teaspoon salt
Freshly ground black pepper to taste
½ cup freshly grated Parmesan cheese

Preheat oven to 325 degrees. Butter a large baking dish. Wash the broccoli and separate, cutting off the stalk ends. Place in a large saucepan filled with boiling water and boil for 4 minutes or until barely tender. Remove with a slotted spoon and place in the baking dish. Meanwhile, melt the butter in a small, heavy saucepan. Slowly whisk in the flour, simmering for 1 minute. Pour in the cream and beef stock and whisk until combined. Cook over moderate heat for 5 minutes, stirring occasionally. Remove from heat and gradually stir in the wine, lemon juice, salt, and pepper. Pour this mixture over broccoli in baking dish and sprinkle cheese on top. Bake in oven for 10 to 15 minutes.

BROCCOLI WITH LEMON AND BACON

Serves 6

1½ pounds fresh broccoli
4 slices bacon
¼ cup **lemon** juice
Salt and freshly ground black pepper to taste

Wash the broccoli, separate flowerets, and cut off the stalk ends. Place in a large saucepan filled with boiling water and boil for 8 minutes or until tender. Meanwhile, cut the bacon in ½-inch pieces and fry until crisp. Add lemon juice to bacon and heat over low heat for 3 minutes. Drain broccoli and arrange on a serving platter. Pour bacon and lemon over broccoli, season, and serve.

DILLED LIMA BEANS

Serves 6

1½ pounds shelled fresh lima beans
1 teaspoon salt
2 tablespoons butter
2 tablespoons olive oil
3 tablespoons **lemon** juice
1½ tablespoons finely chopped fresh dill weed
Coarse salt and freshly ground black pepper

Put lima beans, salt, and butter in a large saucepan. Add enough cold water to cover the beans. Bring to a boil and cover. Simmer for 25 minutes, then drain. Mix olive oil, lemon juice, and dill, and pour over beans. Toss them well—until all the beans are coated and glossy. Season to taste with salt and pepper and serve.

BERNADETTE'S ASPARAGUS

Serves 6

2½ pounds asparagus
1 teaspoon salt
3 tablespoons olive oil
½ cup **lemon** juice
Salt and freshly ground black pepper

Wash asparagus and cut off the tough ends. Place stalks in a saucepan, adding enough cold water to cover them. Add the salt. Bring to a boil and simmer, covered, for 10 minutes or until bottoms are tender. Remove the asparagus carefully with tongs and drain well. Place stalks on a heated platter and pour the olive oil over them, rolling them gently so that they are well coated. Pour lemon juice over asparagus and repeat the rolling until the mixture is blended and the stalks are well coated. Salt and pepper to taste and serve immediately.

RED CABBAGE

Serves 8

2 medium heads red cabbage
3 tablespoons butter
1 tablespoon molasses
2 apples, peeled, cored, and sliced
1 small onion, grated
¼ cup lemon juice
Salt and freshly ground black pepper to taste
½ cup red wine

Wash and shred cabbage. Melt the butter in a heavy skillet and add the cabbage and molasses. Brown over low heat, stirring occasionally. Add the apples, onion, lemon juice, salt, and pepper, and continue stirring. Cover and simmer for 2 hours. Add wine and cook for 30 minutes. Remove from heat and serve.

CAROTTES RÂPÉES

Serves 6

12 medium carrots
¼ cup olive oil
2 tablespoons lemon juice
1 shallot, minced
Pinch of salt
2 tablespoons minced fresh parsley
½ cup Mediterranean-style black olives

The secret of excellent carottes râpées is in the size of the carrot pieces. Ideally, the pieces should be about ½ inch long and thicker than a shaved or scraped piece so that the pieces are crunchy and firm. The standard shredding blade on a food processor makes a good râpé, or you can use a Mouli-type grater.

Scrape the carrots. Cut off the extreme tips and the ends. Grate the carrots, and set aside. Vigorously mix the oil, lemon juice, shallot, salt, and parsley. Pour over the carrots and toss. Mound the carrots on a dish and surround with olives.

SPROUTS À LA BRIGOULE

Serves 6

1½ pounds fresh Brussels sprouts
1 teaspoon salt
1½ cups beef stock
½ cup chopped carrots
¼ cup chopped onions
½ cup chopped cooked chestnuts
¼ cup chopped celery
3 tablespoons butter, cut into small bits
Three ¼-inch slices **lemon**, quartered
Salt and freshly ground black pepper to taste

Preheat oven to 350 degrees. Wash sprouts thoroughly in cold water. Place in a large kettle of water with 1 teaspoon salt and bring to a boil. Cover and boil for 7 minutes or until tender. Grease a medium-size casserole dish. Drain sprouts on paper towels and place in casserole dish. Pour stock into a saucepan and add carrots, onions, chestnuts, and celery, and cook over medium heat for 10 minutes. Remove from heat and stir in butter. Add lemon, salt, and pepper, and pour over sprouts in casserole. Bake in oven for 20 minutes.

LEMONY CREAMED SPINACH

Serves 8

4 slices bacon, cut in 1-inch pieces
2 tablespoons butter
1 medium onion, finely chopped
1 small garlic clove, minced
3 tablespoons flour
1¾ cups milk
2 packages (10 ounces each) frozen chopped spinach, cooked, well drained
2 tablespoons grated Parmesan cheese
1 teaspoon **lemon** zest
1 tablespoon fresh **lemon** juice
½ teaspoon salt
Freshly ground black pepper to taste

In skillet, cook bacon until crisp. Remove and drain. Pour off all but 1 tablespoon drippings. Add butter to same skillet, and sauté onion and garlic in drippings and butter until tender. Blend in flour. Cook a few minutes, stirring constantly, Remove from heat; gradually stir in milk. Return to stove and stir over medium heat until thickened. Add spinach, Parmesan cheese, lemon zest and juice, cooked bacon, salt, and pepper; cook over low heat for 10 minutes. Serve immediately.

MUSHROOMS AND ENDIVES

Serves 8

1 pound fresh mushrooms
½ pound Belgian endives
2 tablespoons **lemon** juice
½ cup olive oil
1 teaspoon salt
Freshly ground black pepper to taste

Wash and dry the mushrooms. With a sharp knife, slice them into thin pieces. Cut endives in half, separate, wash, dry, and add to mushrooms in a large bowl. Pour the lemon juice over all and toss well. Add olive oil, salt, and pepper, and mix gently until well blended. Chill for 2 hours and serve.

SEASONAL CARROTS

Serves 4

2½ cups ½-inch-thick carrot slices
2 tablespoons butter
2 tablespoons brown sugar
1 teaspoon salt
3 teaspoons **lemon** juice
1 teaspoon finely chopped fresh basil
⅛ teaspoon ground paprika
1 teaspoon finely chopped chives
Freshly ground black pepper to taste

Place carrots in a heavy saucepan with enough water to cover them. Bring to a boil and add the butter, brown sugar, salt, lemon juice, and basil. Cover and boil until the water evaporates and the carrots begin to brown lightly—about 5 minutes. Remove to a platter and sprinkle the paprika, chives, and pepper over all and serve immediately.

STUFFED ZUCCHINI

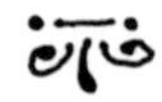

Serves 6

6 small zucchini
1 teaspoon salt
Juice of ½ medium **lemon**
3 tablespoons olive oil
¼ pound mushrooms, chopped
1 small garlic clove, minced
¼ cup chopped fresh parsley
Salt and freshly ground black pepper to taste
½ cup grated Parmesan cheese

Preheat oven to 375 degrees. Drop the zucchini into boiling water for 6 minutes. Drain. Cut zucchini in half lengthwise, and scoop out the centers—be careful not to break the skins. Sprinkle the skins with salt and lemon juice.

Heat the oil in a frying pan, and add the mushrooms and garlic. Cook over moderately hot heat until mushrooms are tender, then add the scooped-out zucchini and cook a few more minutes, stirring. Mix in the parsley, salt, and pepper. Stuff the mixture into the zucchini skins. Sprinkle with Parmesan, put into the oven, and heat until browned.

LEMON RICE

Serves 6

2½ cups water
1 cup long-grain rice
¼ cup butter
½ teaspoon salt
½ teaspoon mustard seed
1 teaspoon turmeric
¼ cup **lemon** juice

Bring the water to a good boil over high heat, drop in rice, cover, and reduce heat to low. Cook for 20 minutes or until the grains are tender. Remove from heat. In a small frying pan, melt the butter. Add the salt, mustard seed, and turmeric. Stir until blended and heat until the seeds begin to pop. Add the lemon juice, give a quick stir, and pour mixture over the rice. Mix well and serve immediately.

HOT BUTTERED CUCUMBERS

Serves 6

3 medium cucumbers
¼ cup butter
Salt and freshly ground black pepper
1 tablespoon **lemon** juice
¼ cup chopped fresh parsley
1 tablespoon chopped chives

Peel the cucumbers, slice them in half lengthwise, scoop out the seeds, and cut cucumbers into ¼-inch slices. Drop the cucumbers into salted boiling water, cover, and simmer for 3 minutes. Drain.

Melt the butter over low heat in a large casserole. Add the cucumbers and sprinkle with salt and pepper to taste. Cook over low heat for 15 minutes; then sprinkle the lemon juice, parsley, and chives over all. Serve while still warm.

BRAISED FENNEL

Serves 6

6 small fennel bulbs or hearts
6 tablespoons butter
1 tablespoon **lemon** juice
¼ cup minced fresh parsley
Salt and freshly ground black pepper to taste
¼ cup minced scallions

Trim the fennel of leaves and strip off the tough, outer stalks. Rinse thoroughly. Place the fennel bulbs in a single layer in a heavy casserole dish, pour in enough boiling water to cover by about 1 inch, and boil, partially covered, for 30 minutes, or until fennel is tender. Drain. Slice the bulbs lengthwise into two or three slices.

Over low heat melt the butter in a large frying pan; add the fennel, lemon juice, parsley, salt, and pepper. Turn the fennel in the butter until it is well coated. Scatter scallions over all and serve.

SAUCES

SAUCES

We begin with Hollandaise. Nothing more than an egg-thickened, lemon-flavored sauce, Hollandaise has become a symbol of haute cuisine partly because it is so tricky to make. One lapse, and your satin-smooth sauce is just scrambled eggs. Hollandaise is inevitably associated with asparagus or broccoli; but if you're proficient at making it, show off and serve it with fish and chicken, too.

Mayonnaise, another egg-thickened sauce, is less difficult—particularly with our recipe, which is the foolproof blender variety. Here, too, if you make a creditable Mayonnaise, don't be shy about serving it with all types of cold foods, not merely salads.

Lemon Butter Sauce is also called Meunière because it's most often put on fish. But the French have given it a special additional role: Lemon Butter Sauce dabbed on grilled steak just at the last moment before serving. When used that way, Lemon Butter Sauce becomes Maître d'Hôtel, as in Contrefilet à la Maître d'Hôtel.

Two sauces which will embellish your leftovers are Piquant Green Sauce and Walnut and Horseradish Sauce. Use them with cold meat, chicken, or fish; you might also try Piquant Green Sauce with leftover vegetables.

HOLLANDAISE

Makes about 1 cup

½ pound butter
2 egg yolks
2 tablespoons **lemon** juice
⅛ teaspoon salt
⅛ teaspoon cayenne pepper

In a double boiler, preferably glass, heat butter until frothy. Keep warm over low heat. Put the egg yolks and lemon juice in a large bowl and beat, on low speed, with an electric beater. Continue beating while adding salt and cayenne. Now add the butter, a few drops at a time in order to keep the sauce from curdling. Keep warm until ready to serve. If the hollandaise does curdle, a few seconds in your blender should bring it back to life.

LEMON SESAME DRESSING

Makes about 1 cup

⅔ cup salad oil
Juice of 1 fresh **lemon**
2 tablespoons vinegar
2 tablespoons toasted sesame seeds
1 tablespoon sugar
1 teaspoon finely chopped onion
½ teaspoon salt

In jar with lid, combine all ingredients; shake well.

EGG SAUCE

Makes about ¾ cup

1 egg
1 finely chopped shallot
1 tablespoon chopped fresh parsley
3 tablespoons olive oil
1 tablespoon **lemon** juice
¼ teaspoon chopped fresh chives
1 teaspoon prepared Dijon-type mustard
Salt and freshly ground black pepper to taste

Boil egg over medium heat for 3 minutes. Meanwhile, mix the shallot, parsley, olive oil, lemon juice, chives, mustard, salt, and pepper in a small bowl. Run the egg under cold water for 20 seconds, crack open, and scoop the yolk into the sauce. Stir for 1 minute. Scoop the white into a plate and chop into small pieces. Add to sauce and stir once. Serve with steak or fish.

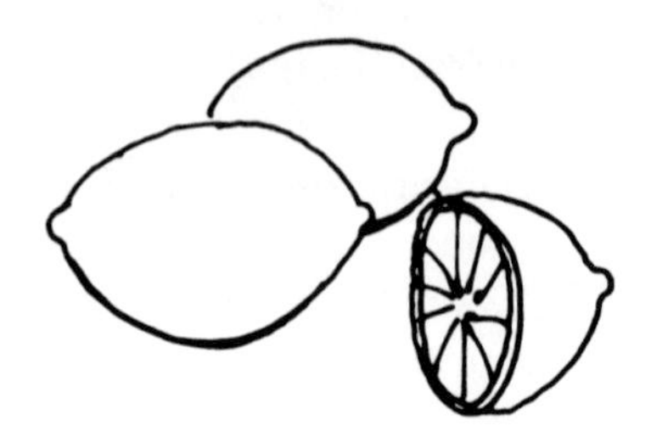

LEMONY APPLE SAUCE

Makes about 1½ cups

3 medium apples, peeled, cored, and quartered
¼ cup **lemon** juice
5 tablespoons chicken stock
Dash of ground cinnamon
Salt and freshly ground black pepper to taste

Cook the apple pieces in the lemon juice for 15 minutes in a covered saucepan over a low heat. Remove from heat and let cool. Combine the chicken stock, apples with cooking liquid, cinnamon, salt, and pepper in a blender or food processor, blending until mixture is smooth. Pour into dish and serve cold or hot with chicken.

PIQUANT GREEN SAUCE

Makes about ½ cup

3 tablespoons chopped shallots
2 anchovies, drained and chopped
3 tablespoons capers, drained and mashed
3 tablespoons minced fresh chives
¼ cup chopped fresh parsley
2 garlic cloves, finely chopped
3 tablespoons **lemon** juice
¼ cup olive oil
Salt and freshly ground black pepper to taste

Mix the shallots, anchovies, capers, chives, parsley, and garlic in a large bowl. Stir in the lemon juice and olive oil. Add salt and pepper. Whisk for 2 minutes, or immediately before serving. This sauce is excellent with a cold veal roast.

TOM'S MAYONNAISE

Makes about 1½ cups

1 whole egg
2 tablespoons **lemon** juice
1 cup vegetable oil
¼ teaspoon salt
½ teaspoon dry mustard

Combine egg, lemon juice, ¼ cup of oil, salt, and dry mustard in a blender and mix for 10 seconds. While blender is on medium speed, very slowly add the remaining ¾ cup of oil. Stop blender as soon as mayonnaise thickens.

LEMON BUTTER SAUCE

Makes ¾ cup

½ cup butter
½ teaspoon salt
¼ teaspoon freshly ground black pepper
2 tablespoons finely chopped fresh parsley
1 tablespoon **lemon** juice

Keep butter at room temperature until the consistency is soft. Beat in the salt, pepper, and parsley. Stir in, a drop at a time, the lemon juice. Serve either chilled or at room temperature with steak or put a pat of this on a broiled hamburger.

WALNUT AND HORSERADISH SAUCE

Makes 2 cups

½ cup heavy cream
2 ounces shelled and skinned walnuts
2 tablespoons freshly grated horseradish
1 teaspoon prepared Dijon-type mustard
½ teaspoon sugar
3 tablespoons **lemon** juice
Salt

Whip cream until it is thick, but stop before it is able to form peaks. Chop walnuts fine by hand or in a food processor. Fold walnuts into the cream. Stir in the horseradish, mustard, and sugar. Slowly blend in the lemon juice. Add salt to taste. Serve with cold roast beef or ham.

MUSTARD SAUCE

Makes about 2 cups

3 tablespoons butter
1 tablespoon coarsely chopped shallots
3 tablespoons flour
1 cup fish stock or substitute clam broth
½ cup dry white wine
1½ tablespoons lemon juice
1 teaspoon dry mustard
2 egg yolks
Dash of ground nutmeg
Salt and freshly ground black pepper

Melt the butter in a medium-size saucepan. Add the shallots, stirring them for 1 minute. With a slotted spoon, remove and discard the shallots. Slowly add the flour, stirring with a fork so that it doesn't lump. When the flour is completely blended, gradually add the fish stock and white wine. Cook over low heat for 3 minutes or until the sauce thickens. Stir in the lemon juice and dry mustard; remove from heat. Just before serving, add the egg yolks, beating in slowly. Return to heat for 1 minute, add nutmeg, and season to taste with salt and pepper. Excellent with fish.

DESSERTS

DESSERTS

Lemon ranks with chocolate as a principal dessert flavor—it's found in ice cream, pudding, cake, pie, soufflé, cookies, sherbet, and candy. Lemon actually has a slight edge over chocolate after a rich dinner; a chocolate dessert might deck the diner, but a plate of something redolent of lemon seems to decalorize all that came before. Indeed, it's that almost mystically perky property which makes lemon first choice for ices served entr'acte during six-course dinners.

Lemon desserts are excellent finishes for opulent meals and summer menus. For example, cold Lemon Pudding, Citrus Ice, and Raspberry Sherbet have a light touch calorically and a refreshing taste and temperature.

Butter Cookies and Lemon Cookies are tart enough to be served with a richer dish, like chocolate ice cream or mousse. Apricot Leaves—with jam and almond batter—are extravagant enough by themselves, perhaps with a strong demitasse.

If you're having fresh fruit that isn't especially fresh—and looks it—pour on some Lemon Sauce for an immediate improvement. This trick works surprisingly well with strawberries and blueberries.

LEMON LOAF

Serves 8

2 cups unsifted all-purpose flour
2 teaspoons baking powder
¼ teaspoon salt
½ cup butter, softened
1 cup sugar
2 eggs, beaten
½ cup milk
1 tablespoon grated **lemon** zest
⅓ cup chopped walnuts

TOPPING

⅓ cup superfine sugar
⅓ cup **lemon** juice
½ teaspoon grated **lemon** zest

Preheat oven to 350 degrees. Grease a medium-size bread pan lightly. Sift together flour, baking powder, and salt. Combine butter with sugar until well blended. Add eggs, one at a time, beating well. To this add half of the milk, then half of the flour mixture, and repeat, beating until combined. Stir in 1 tablespoon lemon zest and walnuts, and pour in bread pan. Bake for 1 hour—a fork inserted in center should come out clean. Meanwhile, mix superfine sugar, lemon juice, and zest in a saucepan and heat for 2 minutes. When bread is removed from oven, immediately pour topping on bread. Let cool for 20 minutes in pan. Remove to rack. Let cool.

LEMON MERINGUE PIE

Makes one 9-inch pie

1½ cups sugar
¼ cup plus 2 tablespoons cornstarch
¼ teaspoon salt
½ cup cold water
½ cup **lemon** juice
3 egg yolks, well beaten
2 tablespoons butter or margarine
1½ cups boiling water
1 teaspoon **lemon** zest

1 baked 9-inch pastry shell

MERINGUE

3 egg whites, at room temperature
¼ teaspoon cream of tartar
6 tablespoons sugar

Preheat oven to 350 degrees. In a 2- to 3-quart saucepan, mix 1½ cups sugar, all the cornstarch, and salt together, using a wire whisk. Gradually blend in the cold water, then lemon juice, whisking until smooth. Add beaten egg yolks, blending very thoroughly. Add butter or margarine. Add the boiling water gradually, stirring constantly with rubber spatula. Gradually bring mixture to full boil, stirring gently and constantly with spatula over medium to high heat. Reduce heat slightly as mixture begins to thicken. Boil slowly for 1 minute. Remove from heat and stir in lemon zest. Pour hot filling into baked pastry shell. Let stand, allowing a thin film to form while preparing meringue.

Use a small deep bowl when beating egg whites. Beat with an electric mixer several seconds until frothy. Add cream of tartar. Beat on high speed until whites form soft peaks. Reduce speed to medium while adding 6 tablespoons sugar gradually, about a tablespoon at a time. Return to high speed and beat until whites are fairly stiff but still glossy and soft peaks are again formed when beaters are withdrawn. Place meringue on the hot filling in several mounds around edge of pie. Using narrow spatula, push meringue gently against inner edge of piecrust, sealing well. Cover the rest of the filling by spreading the meringue from edge of pie to center, making some artful swirls with spatula. Bake for 12 to 15 minutes, until golden brown. Cool on wire rack at room temperature for 2 hours before cutting and serving. Use sharp knife and dip into hot water after each cut for a clean slice.

LEMON ICE CREAM

Serves 6

1 cup sugar
2 tablespoons **lemon** juice
1½ tablespoons grated **lemon** zest
1 pint heavy cream

Combine sugar, lemon juice, and zest, and stir until well blended. Whip the cream until it thickens slightly. Gradually stir in the lemon mixture and pour into freezer tray. Place in freezer compartment, wait for 45 minutes, remove from freezer, and stir lemon mixture, smoothing out any lumps. Return to freezer for 4 hours or until frozen. It's a good idea to let this ice cream soften for 15 minutes before serving.

LEMON PUDDING

Serves 4

½ cup sugar
2 tablespoons flour
½ teaspoon salt
3 egg yolks
1¼ cups milk
3 tablespoons **lemon** juice
2 teaspoons grated **lemon** zest
3 egg whites
½ pint heavy cream

Preheat oven to 350 degrees. Combine sugar, flour, and salt in a large mixing bowl. Beat egg yolks and stir into milk. Add to mixture in bowl. Add lemon juice and zest. Beat egg whites until they form peaks. Gently fold into lemon mixture. Meanwhile, grease a 7- or 9-inch soufflé dish and pour in mixture. Place into pan of hot water and put in oven for 50 minutes. Remove and chill for 4 hours. When you're ready to serve, whip heavy cream until peaks form. Add a dollop of whipped cream to each serving.

HOT LEMON SOUFFLÉ

Serves 6

½ teaspoon powdered sugar
3 tablespoons sifted all-purpose flour
¾ cup milk
⅓ cup granulated sugar
4 egg yolks
1 tablespoon butter
4 egg whites
Pinch of salt
1 tablespoon granulated sugar
¼ teaspoon vanilla
1 teaspoon grated **lemon** zest
¼ cup **lemon** juice

Preheat oven to 400 degrees. Use a buttered charlotte mold or an 8-inch ovenproof baking dish, extended up 4 inches with collar made of foil. Sprinkle powdered sugar over all. Turn mold over to remove excess sugar. In a saucepan, beat flour with a splash of the milk until blended. Beat in the rest of the milk, then mix in ⅓ cup granulated sugar. Place over moderate heat and bring to a boil, stirring vigorously. Let the mixture boil 30 seconds until it becomes very thick. Cool mixture for several minutes, stirring occasionally. Whip in the egg yolks. Beat in the butter. Set aside.

Beat the egg whites with a pinch of salt until soft peaks form. Then sprinkle in 1 tablespoon granulated sugar and the vanilla and beat until stiff peaks form.

Add the lemon zest and juice to the milk mixture. Stir in a large dollop of egg white to lighten the mixture. Then slowly fold the mixture into the egg whites. Pour all into the mold. (To create a "top hat" on the soufflé, run your index finger around the top, ½ inch deep, about 1 inch from the sides of the mold.) Place soufflé in oven, then turn oven down to 375 degrees. Bake for 30 to 35 minutes, until soufflé is puffed and golden. Serve at once.

RASPBERRY SHERBET

Serves 8

Three 10-ounce packages frozen raspberries
1 **lemon**
4 egg whites
Pinch of salt
½ cup powdered sugar
2 tablespoons framboise (optional)

Defrost the raspberries in a strainer set over a bowl. Using the back of a spoon, press the berries until you have 2 cups of raspberry juice. Juice and strain the lemon, then combine it with the raspberry juice. (Save the berry pulp.) Pour the fruit juice mixture into an ice cube tray and freeze for 1 hour. Beat the egg whites with a pinch of salt in a copper bowl until they are frothy white. Add the sugar (you might want to use less sugar if the berries were frozen in sugar syrup). Beat again until egg whites are stiff. Chill for 1 hour. Mix the berry juice, which should be slushy, and the egg whites together.

When well blended, pour into an 8-inch springform pan, and put into freezer for at least 4 hours, or until firmly set. Make the sauce by combining 1 cup (or less) of the raspberry pulp; add the framboise if desired. Chill. When ready to serve, unmold the sherbet and pour the sauce over the top, letting it drip down the sides of the sherbet so that it resembles icing.

LEMON BUTTER DESSERT SAUCE

Makes 1½ cups

⅓ cup butter
1 cup sugar
1½ teaspoons finely grated **lemon** zest
¼ cup **lemon** juice
1 egg, well beaten

Melt the butter in a heavy saucepan and gradually add the sugar, lemon zest, lemon juice, and egg. Cook over moderate heat until just under boiling temperature. Remove from heat and serve hot, with a pound cake, sponge cake, or any sweet quick bread.

LEMON AMANDINE

Serves 8

3 tablespoons flour
½ cup sugar
¼ teaspoon salt
¼ cup butter
4 egg yolks
3 teaspoons grated **lemon** zest
⅓ cup **lemon** juice
1 cup evaporated milk
1 cup cooked rice
5 egg whites
Dash of salt
⅓ cup slivered almonds, toasted

Preheat oven to 350 degrees. Combine flour, sugar, salt, and butter in a large bowl, and cream until fluffy. Beat the egg yolks, one at a time, into the mixture. Gradually add zest, lemon juice, milk, and rice, and stir for 1 minute. In a separate bowl, beat egg whites with dash of salt until they stand in stiff peaks. Fold into lemon mixture and gently blend well. Pour into greased rectangular medium-size baking dish, and place this into pan of hot water. Sprinkle almonds over all and bake for 50 minutes.

LEMON SAUCE

Serves 6

½ cup sugar
1 tablespoon cornstarch
¼ cup salt
1 cup boiling water
1 teaspoon grated **lemon** zest
3 tablespoons **lemon** juice
2½ tablespoons butter

In a heavy saucepan, mix the sugar, cornstarch, and salt and gradually pour in the hot water, blending thoroughly. Bring to a boil and cook 15 minutes, stirring until smooth, thick, and clear. Add the lemon zest, juice, and butter and remove from heat. Stir for another minute or so. Serve immediately.

ANN MARIE'S VERY RICH CHEESECAKE

Serves 8 to 10

CRUST
2 cups graham crackers
¼ cup butter
1 teaspoon cinnamon
¼ cup sugar

CHEESECAKE

Zest of 1 **lemon**
32 ounces cream cheese (room temperature)
1½ cups granulated sugar
¼ teaspoon salt
8 eggs

TOPPING

2 cups sour cream
1 teaspoon vanilla
3 teaspoons sugar
Lemon zest curls

Preheat oven to 350 degrees. Grease the bottom and sides of a cheesecake pan or springform pan. In a large bowl, mix the graham crackers, butter, cinnamin, and sugar until well blended. With your hands, press this mixture down evenly in the bottom of pan and 2 inches up the sides of the pan. Grate the lemon into the cream cheese (reserving a little for the topping) and beat with an electric mixer at a medium speed until very creamy. Stir in the sugar, salt, and eggs and continue beating at medium speed for 10 minutes. (This is the secret to a creamy cheesecake.) Pour into crumb-lined pan and place in center of rack in preheated oven for 65 minutes or until cake is set. Remove from oven and let cool for 20 minutes. In the meantime, beat the sour cream, sugar, and vanilla until the sugar is dissolved. Pour the mixture over the cake and spread with a spatula. Return to oven (350 degrees) and bake for 10 minutes. Chill well before serving and then garnish with a few lemon zest curls.

CITRUS ICE

Serves 8

¾ pound sugar cubes
4 oranges
2¼ cups water
⅓ cup **lemon** juice
⅛ teaspoon salt

Rub a handful of the sugar cubes over the skins of the oranges briskly, until they are saturated with the oil. Then squeeze the oranges and reserve the juice. Bring the sugar cubes to a boil in the water over a high heat and continue boiling for 8 minutes. Remove from heat and let cool. When room temperature, stir in the orange juice, lemon juice, and salt. Pour into an ice cube tray and place in freezer for 5 hours. Be sure to stir every hour while ice is in freezer. Serve in individual bowls with Butter Cookies or Lemon Cookies.

WHITE LEMON FUDGE

Makes about 1 pound

2 tablespoons butter
2 cups sugar
¾ cup milk
1 tablespoon grated **lemon** zest
½ cup chopped walnuts

In saucepan, melt butter over medium-low heat. Add sugar and milk; stir until sugar dissolves. Cover; boil 1 minute. Cook, uncovered, without stirring, to soft-ball stage (239 degrees). Cool to lukewarm (about 45 minutes) ; do not stir. Stir in lemon zest. Beat vigorously until mixture loses glossiness and is ready to set; add walnuts. Pour immediately onto waxed paper. Cool. Cut into squares.

LEMON CREAM

Serves 6

1 envelope (1 tablespoon) unflavored gelatin
1/3 cup water
4 egg yolks
1/2 cup sugar
1/2 cup white wine
1/3 cup **lemon** juice
5 teaspoons finely grated **lemon** zest
1/2 cup heavy cream
4 egg whites

Dissolve the gelatin in the water. In a large bowl, beat the egg yolks and immediately add the sugar, white wine, and lemon juice, beating for 2 minutes more. Stir in the gelatin and 3 teaspoons of the zest and set aside. Meanwhile, beat the cream until it forms soft peaks. Fold cream gently into lemon mixture. Beat egg whites until fluffy, and fold them gently into the cream. Refrigerate for 4 hours. To serve, scoop into individual bowls and sprinkle remaining lemon zest over all.

CRÊPES SUZETTE

Serves 6

CRÊPES

¾ cup water
¾ cup milk
3 egg yolks
3 tablespoons Cognac
1 tablespoon sugar
⅛ teaspoon grated **lemon** zest
1½ cups all-purpose flour
5 tablespoons butter, melted

Put all the ingredients in a blender jar, cover, and blend at top speed for a minute—check to make sure all the flour has been incorporated. Refrigerate in a covered jar for at least 2 hours.

When ready to make crêpes, set a plate on a large dishtowel near the stove. Heat a French crêpe or omelet pan over a moderately hot burner. Coat the bottom of the pan with a bit of butter. When pan is hot, pour or ladle in 2 to 4 tablespoons (depending on the size of the pan) of batter. Quickly tilt the pan so that the batter spreads evenly over the bottom. Cook for a minute or two, or until underside is light brown. Flip the crêpe, using your hands or a spatula or, if you're very adept, by tossing it over with the pan. Brown for another 30 seconds. Slide the crêpe onto the plate resting on the dishtowel. Proceed with the rest of the batter, stacking the crêpes on the plate as you go along. When all the crêpes are cooked, fold the towel over them until the sauce is made and you are ready to serve. If you make the crêpes several hours in advance, reheat them for 10 minutes in a 350-degree oven.

SAUCE

5 sugar lumps
1 medium orange
4 tablespoons unsalted butter
1 teaspoon **lemon** juice
½ cup Grand Marnier
¼ cup Cognac

Rub the sugar lumps all over the orange, then put them on a cutting board and crush them. Scrape the crushed sugar into a chafing dish. Cut the orange in quarters, remove the seeds, and squeeze the juice into the chafing dish. Discard the rind. Add the butter and lemon juice and mix well. Place the chafing dish over the heat and cook until the sugar and butter have melted and start to bubble. Then, using a spoon and a fork, take a crêpe, dip it into the sauce, and fold it twice (into a wedge). Move it to the edge of the dish. Continue until all the crêpes have been folded. Pour the Grand Marnier and Cognac over the crêpes. Light a match and swiftly ignite the liquor in the dish —but do avert your face to avoid contact with the dramatic *pouffff!* Spoon the flaming sauce over the crêpes. When the fire dies out, serve the crêpes to your spellbound guests.

BUTTER COOKIES

Makes 4 dozen

½ cup unsalted butter
⅔ cup sugar
4 egg yolks
1 teaspoon finely grated **lemon** zest
2 cups flour
1 teaspoon vanilla extract

Store butter at room temperature until soft. Blend the butter and sugar together in a large bowl until fluffy. Beat the egg yolks into the mixture one at a time, and add the lemon zest. Slowly stir the flour and vanilla into the mixture. Shape this mixture into a long roll—2 inches in diameter—wrap in wax paper, and refrigerate for 45 minutes. Preheat oven to 400 degrees. Grease a cookie sheet with butter. To shape cookies, slice into ⅓-inch-thick slices. Gently roll the slices between your fingers and then pat down into elongated ovals about 3 inches long. Arrange on cookie sheet. Bake for 10 to 12 minutes, or until edges turn brown. Let cool and remove to a dish to serve.

MADELEINES

Serves 6

4 eggs
½ cup sugar
1¼ cups flour, sifted
½ cup butter, at room temperature
Zest of 1 **lemon,** finely grated
½ teaspoon vanilla extract

Preheat oven to 350 degrees. Beat eggs lightly and place in top of double boiler. With a wooden spoon, stir in the sugar and blend over low heat until mixture is creamy and lukewarm. Remove from heat and continue stirring until it is cool. Gradually beat in the flour, butter, lemon zest, and vanilla. Butter and lightly flour small shell-shaped madeleine molds and pour batter into each until they are two-thirds full. Bake in oven for 20 minutes or until very light golden color. Let cool and remove to a dish and serve.

LEMON COOKIES

Makes 4 dozen

1 lemon
1 egg
½ cup butter, at room temperature
1⅛ cups granulated sugar
1½ tablespoons sour cream
1½ cups flour
¼ teaspoon baking soda
¾ teaspoon baking powder
¼ cup confectioners' sugar
1 tablespoon **lemon** juice

Preheat oven to 350 degrees. Chop the lemons into small pieces and place in food processor or blender. Add the egg, butter, granulated sugar, and sour cream, and mix well. Sift the flour, baking soda, and baking powder into a large bowl. Gradually mix the lemon mixture into this, blending well. Mix the confectioners' sugar and lemon juice in a separate bowl and set aside. Drop small spoonfuls of cookie batter onto greased cookie sheet and bake for 20 minutes or until the edges are brown. Remove from oven and cool. Spread a drop of the sugar mixture on each and remove from pan. Set aside to cool.

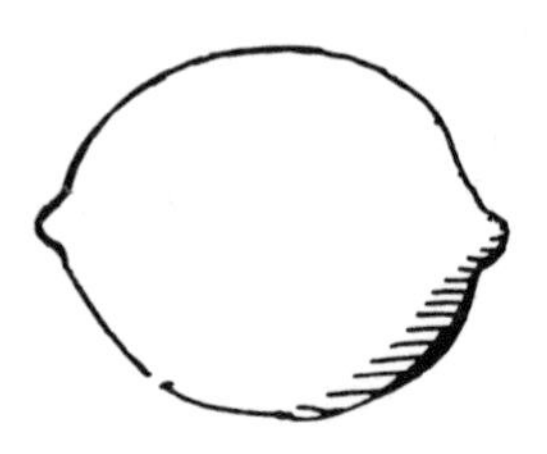

APRICOT LEAVES

Makes 4 dozen

1 cup butter, at room temperature
½ cup sugar
2 egg yolks
¾ cup almond paste
2 tablespoons **lemon** juice
3½ tablespoons grated **lemon** zest
½ teaspoon salt
2¾ cups flour, sifted
1 egg white
Pinch of salt
½ pint apricot jam

Preheat oven to 375 degrees. Cream the butter and the sugar together in a bowl and beat in the egg yolks. Mash in the almond paste and lemon juice; add the zest and the salt. Mix well and slowly add the flour. Form into a ball and refrigerate the dough for 2 hours. Grease and flour a cookie sheet. Sprinkle flour on a large piece of waxed paper and roll out dough until ⅛ inch thick. Set aside extra dough to repeat this. Cut the cookies with a cookie cutter and place on baking sheet. Beat the egg white with the pinch of salt and lightly brush on each cookie before baking. Bake for 12 minutes or until light golden brown. Immediately after removing from oven, spread each cookie with a drop of jam. Let cool and serve.

RYE
GIN
RUM
DRINKS

DRINKS

A bartender without a lemon seems somehow inadequate to life's tasks. Indeed, whole families of drinks require lemon—daiquiris, collinses, sours, and rickeys. Even that drink of drinks, the martini, depends on lemon. (Some authorities frown on a martini which consists of anything but gin and a hint of vermouth, but our recipe is the classic presentation.)

We've also included two medicinal drinks. We do not claim that Peter's Magic Tea will cure everything from ingrown nails to hangover, but it will certainly keep the patient quiet. Bee's Knees is a first-rate elixir for razorblade throat or cough.

And finally, we end our valentine to the lemon with a recipe for—what else?—Lemonade.

JIM'S MARTINI

Serves 6

12 jiggers excellent gin
1 jigger dry vermouth
6 spiral-shaped **lemon** zests

Mix the gin and vermouth in a cocktail shaker with 2 or 3 ice cubes. Pour the martini mixture into 6 iced cocktail or wine glasses. Drop a lemon zest into each glass.

BACARDI

Serves 6

6 jiggers Bacardi rum
3 jiggers **lemon** juice
3 jiggers grenadine

Pour all ingredients in a cocktail shaker with 5 or 6 cracked ice cubes, and shake for a minute. Strain and serve straight up in chilled cocktail or wine glasses.

DAIQUIRI

Serves 6

8 jiggers rum
2 jiggers **lemon** juice
2 teaspoons powdered sugar

Pour all ingredients in a cocktail shaker with 5 or 6 cracked ice cubes, and shake for a minute. Strain and serve straight up in chilled cocktail or wine glasses.

TOM COLLINS

Serves 6

6 jiggers gin
6 **lemons**
3 tablespoons powdered sugar
24 ounces club soda

Use highball glasses. Fill each glass with ice cubes, then stir in a jigger of gin, juice of 1 lemon, 1 teaspoon of powdered sugar, and 4 ounces of club soda.

GIN RICKEY

Serves 6

6 jiggers gin
3 **lemons**
24 ounces club soda

Use highball glasses. Fill each glass with ice cubes, add a jigger of gin, juice of half a lemon, and 4 ounces of club soda, and stir.

PISCO SOUR

Serves 6

6 tablespoons egg white
3 tablespoons powdered sugar
8 jiggers Pisco
3 tablespoons **lemon** juice

Mix the egg white and sugar in a cocktail shaker. Stir in the Pisco and lemon juice. Add 5 or 6 cracked ice cubes, and shake for a minute. Strain and serve in chilled cocktail or wine glasses.

WHISKEY SOUR

Serves 6

8 jiggers rye
3 jiggers **lemon** juice
2 teaspoons powdered sugar

Pour all ingredients in a cocktail shaker with 5 or 6 cracked ice cubes, and shake for a minute. Strain and serve straight up in chilled cocktail or wine glasses.

COQ ROUGE

Serves 6

6 jiggers rum
3 jiggers gin
3 jiggers Cointreau
3 jiggers **lemon** juice

Pour all ingredients in a cocktail shaker with 5 or 6 cracked ice cubes, and shake for a minute. Strain and serve straight up in chilled cocktail or wine glasses.

SALLY'S COLLINS

Serves 1

½ medium **lemon**
½ medium lime
1½ teaspoons superfine sugar
1½ jiggers light rum
½ cup club soda
1 **lemon** twist

Squeeze the juice of the lemon and lime into a tall glass. Add the sugar and rum and stir until well blended. Add ice cubes and pour in club soda; stir and serve with a twist of lemon.

PETER'S MAGIC TEA

Serves 1 sick person

1 cup strong tea, any standard type
3 ounces dark rum
⅓ medium **lemon**
2 teaspoons sugar

Brew a cup of strong tea. Using a large breakfast cup or mug, add the rum, squeeze in the lemon, and stir in the sugar. Serve immediately.

BEE'S KNEES

Serves 6

8 jiggers gin
2 tablespoons honey
⅓ cup **lemon** juice

Pour all ingredients in a cocktail shaker with 5 or 6 cracked ice cubes, and shake for a minute. Strain and serve straight up in chilled cocktail or wine glasses.

LEMONADE

Serves 6

6 cups water
$\frac{1}{2}$ cup **lemon** juice
1$\frac{1}{4}$ cups sugar
Pinch of salt

Mix all the ingredients together in a big pitcher. Chill for 2 hours. Pour over ice cubes in tall glasses.

Index